ENERGY SCIENCE, ENGINEERING AND TECHNOLOGY



# TO OWN OR LEASE SOLAR INSTALLATIONS

## CONSIDERATIONS FOR DECISION MAKERS

# ENERGY SCIENCE, ENGINEERING AND TECHNOLOGY

Additional books in this series can be found on Nova's website under the Series tab.

Additional e-books in this series can be found on Nova's website under the e-book tab.

ENERGY SCIENCE, ENGINEERING AND TECHNOLOGY

# TO OWN OR LEASE SOLAR INSTALLATIONS

## CONSIDERATIONS FOR DECISION MAKERS

JORDAN GOMEZ
EDITOR

NOVINKA

*New York*

## NOTICE TO THE READER



**Library of Congress Cataloging-in-Publication Data**

ISBN: 978-1-63483-080-5

*Published by Nova Science Publishers, Inc. † New York*

# CONTENTS

# PREFACE

This book examines the tradeoffs among financing methods for businesses installing onsite photovoltaics (PV). The authors present case studies of PV financing strategies used by two large commercial retailers that have deployed substantial U.S. PV capacity: IKEA, which owns its PV, and Staples, which purchases power generated from onsite PV systems through power purchase agreements (PPAs). The book also analyzes the financial considerations that influence any company's choice of PV financing strategy. The goal in this book is to clarify the financial and institutional costs and benefits of financing strategies and to inform other companies that are considering launching or expanding similar PV programs. Furthermore, this book provides a high-level overview of the developing U.S. solar loan product landscape, from both a market and economic perspective. It covers current and potential U.S. solar lending institutions; currently available loan products; loan program structures and post-loan origination options; risks and uncertainties of the solar asset class as it pertains to lenders; and an economic analysis comparing loan products to third party-financed systems in California.

In: To Own or Lease Solar Installations
Editor: Jordan Gomez
ISBN: 978-1-63483-080-5


*Chapter 1*

# TO OWN OR LEASE SOLAR: UNDERSTANDING COMMERCIAL RETAILERS' DECISIONS TO USE ALTERNATIVE FINANCING MODELS*

***David Feldman and Robert Margolis***
National Renewable Energy Laboratory

## EXECUTIVE SUMMARY

This report examines the tradeoffs among financing methods for businesses installing onsite photovoltaics (PV). We present case studies of PV financing strategies used by two large commercial retailers that have deployed substantial U.S. PV capacity: IKEA, which owns its PV, and Staples, which purchases power generated from onsite PV systems through power purchase agreements (PPAs). We also analyze the financial considerations that influence any company's choice of PV financing strategy. Our goal in this report is to clarify the financial and institutional costs and benefits of financing strategies and to inform other companies that are considering launching or expanding similar PV programs.

* This is an edited, reformatted and augmented version of technical report NREL/TP-6A20-63216, issued by the National Renewable Energy Laboratory, December 2014.

IKEA's decision to self-finance its PV systems fits well with its corporate culture. Although the company assumed more risk by owning PV instead of leasing[1] it, IKEA believes that careful process management enables it to reap the benefits of owning long-term, high-quality assets. In contrast, Staples' decision to lease its PV systems through PPAs aligns well with its risk-return preferences. The technical and market expertise of the third-party PV owner has enabled Staples to deploy PV rapidly, and the company is pursuing several strategies for overcoming challenges related to installing PV on leased buildings.

Our modeling quantifies financial drivers that companies like IKEA and Staples consider when choosing a PV financing strategy. If one simply assumes a pre-tax discount rate of 10%, the levelized cost of energy (LCOE) for the modeled self-financed system is approximately 30% lower than the LCOE for the PPA-financed system.[2] However, the attractiveness of the financing methods varies due to different assumptions for cost of capital, different perceptions of PV investment risk, and different perceptions of the risk of using a particular financing method. When the commercial customer's pre-tax discount rate rises to 23%, for example, the LCOE is equivalent under either financing method. If a company assumed a 10% pre-tax discount rate for a PPA and a 23% pre-tax discount rate for self-financing, then the LCOE would be 14% lower using the PPA.

The timeline for customer payback is another consideration. Although, in our model, self-financing provides a better return over the 20-year project lifetime, it is net-cash-flow-negative for about 5-11 years. The PPA, on the other hand, produces positive cash flow quickly and is economically more attractive than the self-financed system for the first 6-14 years. Still, our modeled self-financing is even more attractive over the long term if the PV lifetime is extended to an expected period of 30 years—this assumption lowers the LCOE by 17%. Self-financing might also have advantages over PPAs due to involving fewer parties in the financial transaction and requiring less-expensive electric meters.

That said, PPAs offer potential advantages to some companies. Unlike purchase of a PV system, PPAs are off-balance sheet transactions, which might suit companies unwilling or unable to add the liability of a PV loan onto their balance sheets. In addition, a company can get the full benefits of a PPA with any amount of tax liability whereas self-financing companies must have sufficiently large tax liability to take full advantage of PV tax incentives. Companies using PPAs also might reduce their exposure to risks related to PV underperformance, unanticipated O&M costs, and delays in receiving incentives and grid-interconnection approval.

Our analysis suggests that the most appropriate PV financing option for a particular business depends on the characteristics and circumstances of that business.

# 1. INTRODUCTION

Most distributed solar photovoltaic (PV) production capacity in the United States is installed on commercial buildings and property. In October 2013, the top 25 companies had installed approximately 11% of all distributed U.S. PV capacity (SEIA and Vote Solar 2013; SEIA and GTM Research 2014). This likely results from several factors: businesses have large electricity needs, accounting for 36% of all U.S. electricity sales in 2013 (EIA 2014); they occupy substantial real estate, with an estimated total floor space of 88 billion square feet[3] (EIA 2014); and, unlike many individual consumers, they often can make long-term investments or contractual agreements to produce economic benefits.

One of the key decisions a business must make about PV is how to finance its installation. A company can buy a PV system using debt or its own funds. Alternatively, it can sign a long-term contract (15–25 years) to lease the system or purchase the system's electricity (via a power purchase agreement [PPA]) with a third party that purchases, owns, and operates the PV system on the company's property. This third-party-ownership (TPO) model constitutes a significant market share in many U.S. states. In California and New Jersey, two of the top U.S. commercial markets, TPO systems account for approximately 25%–55% of commercial systems (Figure 1).

Each ownership model—self-ownership and TPO—has its own value proposition, with different benefits and limitations. The cost of capital incurred by the host business and the third party is a major factor. However, other factors can affect the economics as well, including the host's need for tax liability to use federal tax incentives, the additional liability and maintenance costs associated with ownership, the complications of adding assets and liabilities onto the company's balance sheet,[4] and the economic time horizon of decision makers.

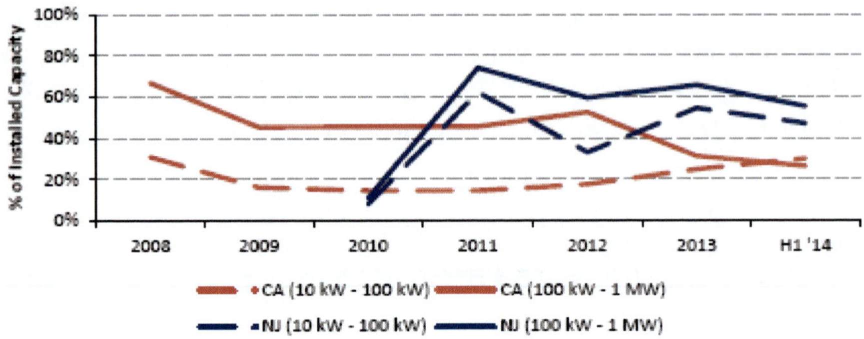


Sources: CSI 2014; NJCEP 2014.

Figure 1. Third-party market share of annual commercial PV installations, by state (California and New Jersey) and system size.

To explore the tradeoffs between PV self-ownership and TPO, this report presents detailed case studies of two companies with contrasting PV-investment strategies: IKEA, a home furnishings retailer that owns all of its systems directly, and Staples, an office-supply chain that purchases power generated from onsite PV systems through PPAs. In 2013, IKEA ranked first in the nation in the percentage of facilities using solar, with 35 MW of PV on 39 (89%) of its U.S. facilities (Figure 2), while Staples ranked ninth in installed PV capacity, with 14 MW of PV on 37 (2%) of its U.S. facilities (SEIA and Vote Solar 2013). At the end of 2013, Staples and IKEA had a combined PV capacity larger than the capacity of 28 individual states (SEIA and GTM Research 2014).

We examine why these two companies chose different ownership and financing structures to achieve significant PV deployment. We also analyze the financial considerations that influence any company's choice of PV financing strategy. Our goal is to clarify the financial and institutional costs and benefits of each strategy from the end user's perspective for the benefit of other companies that are considering launching or expanding similar PV programs. While some of the methods IKEA and Staples pursued were achievable in part because they are multinational companies with large balance sheets, many of the decisions made and lessons learned are applicable to businesses and deployment strategies of many different sizes. Sections 2 and 3 of this report present the IKEA and Staples PV case studies. Section 4 analyzes the economics of PV self-ownership and TPO, and Section 5 provides conclusions.

# 2. IKEA

As of June 2014, IKEA had PV on 40 (nearly 90%) of its U.S. facilities in 20 states, representing a total capacity of 39 MW, as part of a global investment allocation of $1.8 billion in renewable energy (IKEA 2014). Approximately half of the company's global PV capacity currently is in the United States. This large-scale deployment resulted from IKEA's global strategy, which in turn was influenced by the company's structure and corporate culture.

Source: StrataSolar 2014.

Figure 2. IKEA store with substantial PV deployment, Atlanta, GA.

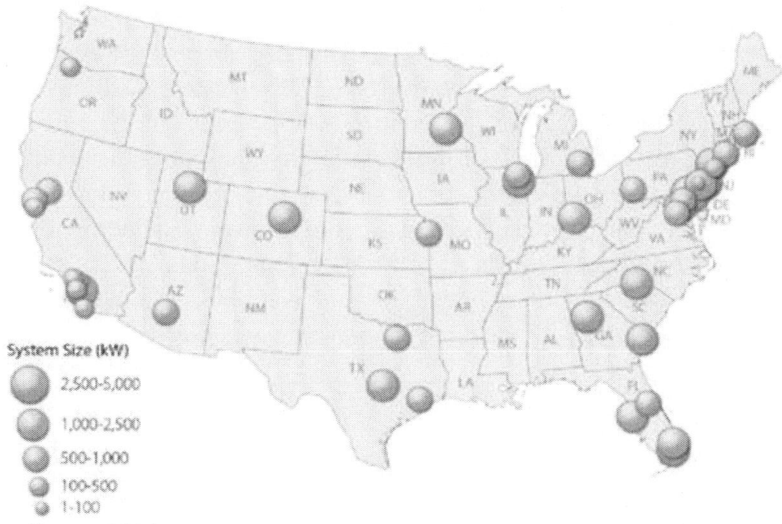


Source: Gasper 2014.

Figure 3. PV Installations at IKEA facilities, by system size.

## 2.1. Deciding to Deploy PV

IKEA is a privately-held company, with many of its business activities driven by global offices in Sweden, but its corporate management delegates certain decisions to local headquarters in individual countries or regions.[5] Around 2005, IKEA's global operations reviewed its business costs, including energy costs. IKEA buildings require substantial electricity, and the company viewed this as an uncontrolled cost. To mitigate energy-related risk, IKEA began developing strategies to reduce energy consumption through energy efficiency and reduce exposure to electricity price changes. It determined that using renewable energy could help control these costs. In particular, PV was viewed as a potential long-term hedge against volatile electricity prices.

Although IKEA does not have to worry about public investor sentiment, it wanted to be a better corporate citizen by reducing its carbon footprint. According to company interviews, costs and sustainability are cornerstones of Swedish culture, and everything IKEA does is filtered through the lens of sustainability (Gasper & Roth 2014). IKEA's private status also enables it to have a longer-term planning horizon compared with the horizons of typical public companies. As IKEA public affairs manager Joseph Roth puts it, the company's mindset is, "How can we reduce costs, and do it in a good way? We are constantly looking for ways to be more sustainable in that context" (Gasper & Roth 2014).

IKEA set a global goal of supporting its operations with 100% renewable energy while reducing energy use. It established a global energy program and delegated the particular strategies for achieving the overarching goal to company leadership in each of its countries of operation. In the early years, IKEA's global team took the lead in identifying opportunities. Thresholds for capital investments were established, and in 2006 the company globally began focusing on PV. Four pilot sites were created worldwide, including a store in the United States (Brooklyn, New York), to evaluate module performance by gathering long-term, granular production and cost data. At the same time, IKEA's U.S. team began learning about the technology and potential frameworks for financing it.

## 2.2. Choosing a Financial Structure

Early on, IKEA determined it had no interest in PPAs for financing PV deployment for several reasons. First, it thought owning the PV would provide a better return. "We had very clear guidance from IKEA globally," says IKEA PV program manager Mark Gasper. "The mindset was that if you look at the PPA model, and what drives companies to be able to get investments to drive that model, there has got to be a reasonable rate of return built into it. If an investor can get a minimum rate of return and can pass some of the savings in the form of an electricity hedge onto the host, then we [IKEA] should be able to cut out the middleman and capture all of the benefits associated with the project, whether those are financial—from a tax perspective, etc.— or environmental attributes (Gasper & Roth 2014).

Second, IKEA believes it operates under a longer investment time horizon than most other organizations are willing to accept. "When IKEA looks at what we consider a minimum of at least 20, or more like a 25-30 year investment, the corresponding payback time –whatever it may be–is not out of line with the time period IKEA expects the asset to be productive for the organization," says Gasper (Gasper & Roth 2014). When IKEA enters a market, it wants to be there a long time, and it typically builds its own buildings, purchases the associated land, and owns or establishes long-term relationships with its suppliers. Roth attributes this desire to own rather than lease to the company's southern Swedish culture. "If you work hard, you can control your own destiny," he says. "Why should we participate in a PPA when we could own it [the PV] ourselves?"

IKEA also can finance capital expenditures through its available balance sheet. It typically tries not to rely on outside financing, whether buying land or buildings or making capital investments. If an investment satisfies a required rate of return and IKEA global approves it, the company can self-finance it.

## 2.3. PV Ownership Challenges and Solutions

Deciding to own PV presented IKEA with a new set of challenges. The company had no experience procuring, owning, or operating a PV system and therefore had no institutional knowledge to understand the process or assess the risks. To overcome these challenges, IKEA began working with external partners that understood their vision and were willing to work with them. PV program leaders at IKEA also engaged the company's finance and tax departments to understand how they could best take advantage of PV ownership.

IKEA's first U.S. project (not driven by the global pilot program) was in Arizona in 2010. The company, which at the time had little knowledge of U.S.

solar markets, partnered with an external engineering, procurement, and construction (EPC) installer who helped guide IKEA through the decision making process. When IKEA's U.S. PV team presented the Arizona proposal, IKEA Global became excited about U.S. PV opportunities. Again partnering with an EPC installer, the U.S. team searched for additional areas with high energy costs and some potential solar incentives and identified a preliminary set of locations that were most financially attractive for PV. With the EPC installer's help, the team enrolled in incentive programs, shifting from Arizona to California and the California Solar Initiative (CSI), which, at the time in 2010, was beginning to reduce its performance-based incentive. Fortunately the team moved fast enough before a flurry of other applications shut down the CSI system for a few weeks later that year. IKEA next analyzed all of its U.S. locations, examining available incentives and local utility rates and then developed a roadmap for the most feasible PV locations. It moved more quickly in locations with high energy costs and available incentives because the team wanted to make sure those opportunities were not missed.

Early on, IKEA wanted to partner with vertically integrated module manufacturers that had a large balance sheet because of their perception that such firms would be less risky than those that were only EPC installers. However, IKEA came to the opinion that modules were a commodity and started focusing on the quality of system installation. Because IKEA would own and manage these systems for 20-30 years, it wanted to minimize maintenance concerns. In addition, roof maintenance is critical to the company, and it was very engaged in the design and construction of its rooftop systems. "We sell wood-based products in flat pack cardboard," says Roth. "If something happens to our roof, we have millions and millions of dollars of inventory at risk." Although IKEA's component and system design requirements were more robust than the requirements of some installers, the company became less stringent on components, especially modules, as long as they met the required specifications. IKEA also tried to use some of the same core equipment throughout its U.S. projects to facilitate O&M.

## 2.4. Expanding Deployment

Initially, IKEA relied on integrators and developers to offer PV opportunities. In a relatively short time, however, the company built an internal U.S. PV team knowledgeable enough to evaluate opportunities and projects on its own, release requests for proposals (RFPs), and work with multiple EPC firms. Releasing RFPs was especially important when entering a new area, where the team needed to identify quality companies that would fit with IKEA's corporate culture. As it gained experience and confidence, the team developed its own expertise in modeling PV opportunities, examining incentives and other factors that drove decision making. Gaining this expertise required commitment over time, but not a large staff; only one to two people worked on the solar projects at a time.

Through its internal analysis, IKEA started developing solar projects in markets with very little solar deployment. This presented challenges, particularly with respect to the interconnection process. Owing to limited experience connecting large PV systems to their networks, some utilities had concerns about the grid reliability impacts of connecting a 1-MW or larger rooftop installation. While some of these discussions took time, IKEA understood how the systems worked and had long-term historical data from other PV systems it owned; in certain cases IKEA put the prospective utilities in contact with utilities that had worked with IKEA. In several states (including Maryland, Minnesota, Georgia, Virginia, Connecticut, and Michigan), the rooftop systems IKEA installed were the largest built at the time, which required additional effort.

IKEA has been able to install several systems that met its hurdle-rate requirements, but which have received no state or local incentives—historically a situation that has been challenging for most companies. Because the company knew the expected system production and electricity rates, it could identify opportunities that met its long-term requirements. While projects with higher expected returns were pursued first, the company has been able to keep identifying viable projects, in part, because of PV system price declines since its PV program began.

As PV owners, IKEA has more O&M responsibilities than those who lease through a third party. Whereas O&M is typically part of a TPO contract, PV owners must maintain systems themselves or contract O&M to a third party. Currently, IKEA has EPC installers providing O&M, managed by IKEA's facilities group. The company still is evaluating best practices for maintaining the systems in the long term, once the initial O&M contracts expire (none have to date).

IKEA had to learn other aspects of managing PV systems as well. In states that used Solar Renewable Energy Certificate (SREC) programs to incentivize solar adoption, IKEA learned and created processes to comply with the necessary administrative procedures. The company also learned how to evaluate PV installation bids. Initially concerned with per-watt system cost, the company came to realize that the per-kilowatt-hour cost was the most important metric. IKEA now requires EPC installers to provide more detail on their system designs and carefully evaluate the expected system production. Over time, the company secured annual system production guarantees for many of its locations to protect itself from underperformance risk.

PV at a few IKEA locations proved financially unviable due to a combination of rates, lack of incentives, and lack of solar radiation, as well as roof-space/ structural limitations that precluded PV installation. Still, the company has installed PV on roughly 90% of its facilities. As shown in Figure 3, IKEA's PV installation efforts peaked in 2012 when it installed 20 MW at 22 of its locations.

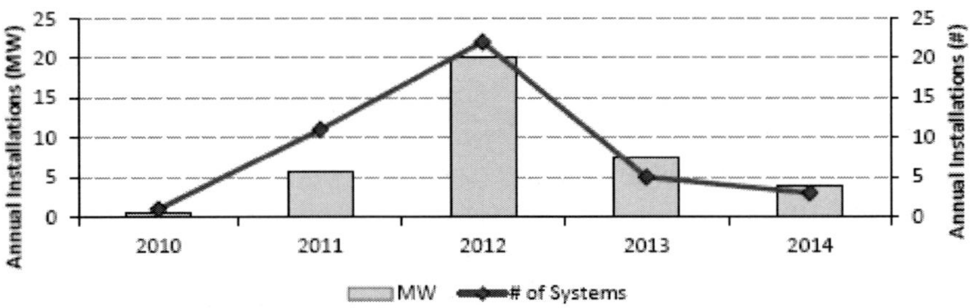


Figure 4. Annual IKEA PV installations, by number of systems and installed capacity.

Today, IKEA feels it has gone as far as it can in using U.S. onsite PV at existing locations, but continues to evaluate solar potential on all future buildings, building on its U.S. solar presence and striving toward a 100% renewable energy goal. The company also is exploring other strategies, such as wind farms–two of which IKEA is currently purchasing in Illinois and Texas. It continues to examine energy management and controls to reduce grid electricity use and the associated demand charges. However, much of its peak daily load occurs from 4:00 to 7:00 p.m., which does not align with a typical PV system's production profile. IKEA is exploring ways of shifting its load curve, but as a retailer it must maintain a standard comfort level for customers. In addition, IKEA's buildings typically consume much more energy than its PV systems can generate. Despite filling entire store rooftops with PV (Figure 2), IKEA still typically consumes every kilowatt-hour the systems produce.[6] Thus IKEA's systems do not typically rely on net-metering agreements to make the economics work.[7]

## 3. Staples

As of the end of 2013, Staples had PV on 37 (2%) of its U.S. facilities in seven states, totaling 14 MW of capacity. It has installed PV on retail stores, distribution centers, and its corporate headquarters. These results are due to the company's corporate strategy, its partnerships, its underlying primary business, and its corporate culture.

Source: CEFIA n.d.

Figure 5. Staples distribution center with substantial PV deployment, Killingly, CT.

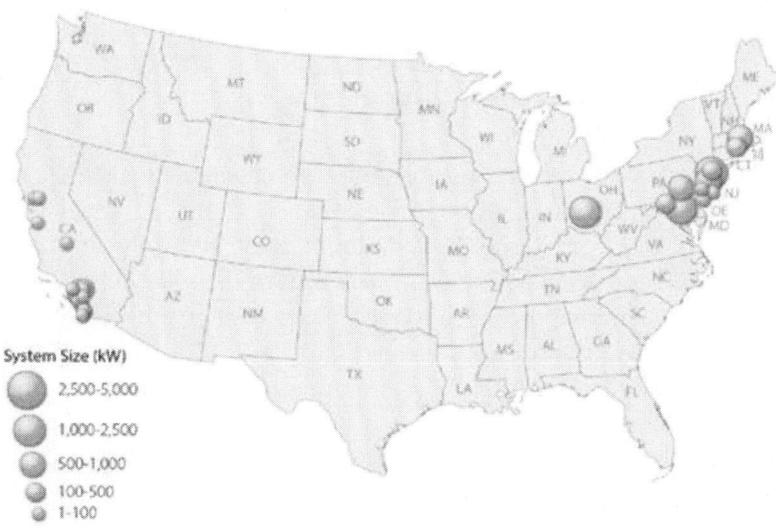


Sources: CSI 2014; CEFIA 2014; DOER 2014; PJM-EIS 2014.

Figure 6. PV Installations at Staples facilities, by system size.

## 3.1. Deciding to Deploy PV

Staples is a 28-year-old company with a diverse portfolio of facilities. At the end of 2013, it operated 1,515 retail stores and 55 distribution centers in 48 states and Washington D.C. as well as 654 retail stores and 61 distribution centers abroad. Facilities range in size from distribution centers with 1 million square feet to city stores with a few thousand square feet.

Around 2002, through the initiative of its chief financial officer (CFO) and vice president of environmental affairs, Staples began to look for energy innovation opportunities beyond the standard practice in the retail sector. Similar companies at the time were focusing primarily on lighting retrofits, which made economic sense. Staples' CFO generated the idea of bundling a portfolio of energy-related retrofits over a large number of locations, even in areas without incentives, to make the whole investment attractive. Beyond the economics, Staples was also interested in becoming a more sustainable company. It instituted a long-term carbon reduction goal and made energy efficiency and renewable energy a key part of fulfilling that goal. It has also been involved in other green initiatives, such as the Green Power Market Development Group, which consists of 15 leading U.S. companies and had, among other goals, a 2010 renewable energy deployment goal. Staples also joined the U.S. Environmental Protection Agency's (EPA) Climate Leaders Program, which it felt encouraged companies to be more sustainable.

## 3.2. Choosing a Financial Structure

Staples viewed PV through sustainability and economic lenses. During its initial examination of PV (2003-2005), the company could not model a case in which PV made economic sense. In addition, 95% of Staples' real estate was leased, thus it did not have control over much of its rooftop space.

In 2003, SunEdison approached Mark Buckley, Staples' vice president of environmental affairs. SunEdison, which was a relatively small company at the time, had been developing a TPO business model to provide solar PPAs to retail and office businesses. It offered to install PV on Staples' property at no cost and sell the system's electricity under a 20-year PPA. The value proposition pitched to Staples was that it did not need to own the asset but, owing to its good credit, could sign the long-term agreement at below-market rates. This proposition intrigued Staples' CFO, John Mahoney Jr., but he had questions. First, why would Staples put assets on a roof it did not own?

Second, given that the average lease term Staples holds is 10 years (with a 5-year option), why would Staples sign an agreement that is longer than its lease?

To solve these issues, Staples and SunEdison modified their PPAs to provide more flexibility for the company. While the length of the PPA was 20 years, they implemented rollover provisions that enabled Staples to exit after the first 5 years without difficulties.[8] After that period, Staples could assign the contract to a different company, such as the subsequent tenant of the building. In addition, if Staples moved out of a building and the new tenant did not want to assume the PPA, SunEdison would relocate a certain amount of PV without any cost to Staples. It was very beneficial to have a portfolio when using this approach because the cost built into the PPA price was spread over many projects. In other words, Staples and SunEdison could balance the potential cost of moving one project against all of Staples' other stores that either did not move or could reassign their PPAs.

With these measures in place and the CFO on board, the Staples PV team had to get buy-in from each of Staples' business departments such as its head of logistics and its head of retail operations. With eventual corporate buy-in, Staples claims they were the first company in the United States to sign a PPA in 2004. Staples did not view purchasing PV systems as an option. It had competing needs for capital and did not think owning PV was part of its core business. PPAs provided the company a hedge against increasing future electricity prices with less risk and responsibility than owning systems outright.

Unlike IKEA, Staples is a public company, which gives it a different perspective on the risks of PV ownership. Staples has more stringent public reporting procedures, thus there was some concern about solar affecting the balance sheet. However, Staples is a very large company with a large balance sheet and a good line of credit; therefore the company's decision regarding solar ownership was primarily associated with the different level of risks. Company personnel hold various views as to the best financing structure, but the current corporate leadership does not have the risk tolerance to be comfortable with PV ownership.

### 3.3. Deploying PV

Because entering into a long-term PPA is not without risk, Staples developed a deployment strategy to maximize its cost savings and minimize its risk. First, although individual system savings varied, PPA rates had to be equal to or less than retail electricity rates. The contracts included escalators ranging from 0.5% to 2.0%, which was below the rate at which Staples thought retail rates would increase. As new deals moved forward, U.S. PV system prices decreased, which were passed along through lower PPA prices to Staples.

Initially, much time was spent getting building owners to allow Staples to put PV on their roofs, but this became easier as landlords became more familiar with the product. Some owners wanted compensation for allowing PV on their roofs, but Staples did not move forward with those locations. The company was concerned about the potential for PV to cause roof leakage, so it worked with SunEdison, the building owners, and roof manufacturers/ installers to ensure that the systems would not damage the roofs or void the warranties.

Early on, Staples was open to using different PV technologies depending on a location's needs. For example, in one location the company installed a thin-film technology due to weight concerns, but their difficulties with the product deterred the company from using the product again. It also installed a large number of ground-mounted arrays, some with tracking technologies, co-located with distribution centers in Maryland, Ohio, and Pennsylvania. However, most of its systems are rooftop installations using standard crystalline-silicon modules. Over time, working in conjunction with SunEdison, the company focused on a set of standard technologies and system designs.

In deciding where to locate the PV systems, Staples and SunEdison looked at all the stores that made economic sense. They focused on states where the cost of energy was high and an RPS or local incentives were in place. The first few stores were built with funding from state grants, and Staples kept the SRECs to fulfill its corporate carbon commitments. However, as states began to reduce upfront subsidies and encourage deployment through SREC programs, the credits became a more important part of financing the installations. In response to this market change, Staples began assigning the SRECs to SunEdison and getting SunEdison to source cheaper RECs from other sources to satisfy its carbon commitments.

Many installation decisions, particularly in the beginning, were handled by SunEdison, which Staples saw as a technical advisor and partner. Staples initially managed the process with one person—its vice president of environmental affairs—but as the portfolio grew, so did the team. SunEdison did most of the market feasibility studies while Staples focused on the building feasibility (e.g., how much weight the building could handle, the roof size and age). Staples also screened all buildings to make sure they had a lease of at least 5-10 years. Collectively, the two companies selected a group of stores that fit these criteria. It was an interactive process. Staples followed trends within the U.S. solar market (such as RPS changes), but it viewed SunEdison as its expert advisor.

Staples has considered other PPA providers and occasionally has released an RFP for a particular installation, but to date it has worked almost exclusively with SunEdison. One of the reasons for this is that Staples values a vertically integrated partner, with expertise and control over installation, operation, and ownership. In the early days of its program, it had heard of third-party system owners contracting out work and the PV systems not working as well as expected (or not completed on schedule). SunEdison now has some strategic partnerships, but this is primarily because they are much larger now, which provides Staples with sufficient confidence in SunEdison's operations.

Staples contracted once with a different company; several years ago, it bought a competitor that was leasing a warehouse. Staples approached the landlord, Hartz Mountain, to put PV on its roof, but the landlord was not interested in allowing a third party to install the system. Instead, the landlord wanted to install the PV itself and offered a PPA coterminous with Staples' lease. Buckley thinks this model makes a lot of sense. "The friction of installing a 100-kW system on a portion of a landlord's roof is so much greater than the landlords doing it themselves for all of their tenants," he says (Buckley 2014).

## 3.4. PPA Challenges and Solutions

Having the landlord own the PV system would address one of the biggest challenges for retail stores, which is that most companies lease rather than own their property. Staples had to negotiate with the landlord of every store on which it wanted to install PV. This problem has been solved to some extent because Staples has added standard language to all its new leases since 2007 to facilitate installing PV. This language stipulates that Staples must let the landlord know it is going to install PV, but the landlord cannot unreasonably restrict Staples.

Staples has recently encountered another large obstacle to further PV deployment: because of changing market conditions, its retail sites are in flux, and the company recently announced 225 store closures. In general, Staples currently does not want to assume the financial risk related to PV generation at its retail locations. The biggest limiting factor for further PV deployment is the company's leased assets. "Unless we know for certain we are not moving out of the location, it is hard to get approval to put on solar," says Buckley (Buckley 2014). Thus Staples' PV installations have plateaued (Figure 5). The company is still committed, however, to installing PV systems where financially attractive. In the current economic environment, this means Staples' focus has shifted to installing PV at its large distribution facilities where it has more stability and owns many of the buildings. It has also completed a few projects in distribution facilities in Belgium and the United Kingdom.[9]

Staples is also pursuing virtual net-metering, where it could get credit for electricity generated by an offsite PV system within its electric utility's service territory. This approach would be particularly advantageous for a company like Staples: it could have a PPA to provide solar electricity to a number of facilities within a region and, if it closed one store, it could simply use the power at a different location within that region. Virtual net-metering would also eliminate the need to negotiate with landlords for use of the roof in retail locations.

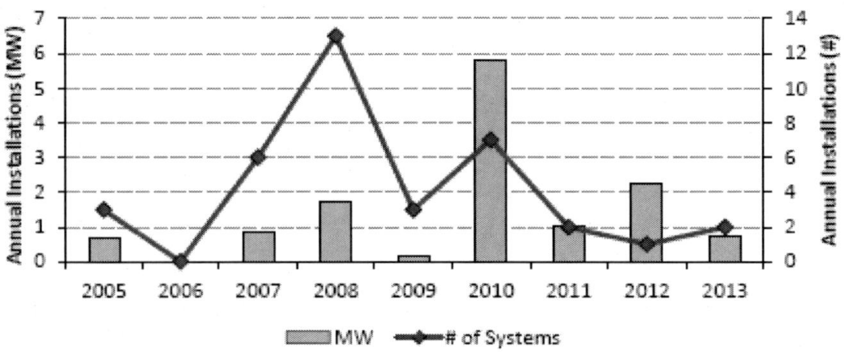


Sources: CSI 2014; CEFIA 2014; DOER 2014; PJM-EIS 2014.

Figure 7. Annual Staples PV installations, by number and capacity.

Despite these concerns, the turnover of stores with PV systems has been relatively small. Only one system has been transferred to a different party (none have been transferred to a different location), and the company reports that the flip was very smooth to the next tenant, who was creditworthy. If the location had remained vacant or the tenant had not been creditworthy, there might have been a problem, or the system would have been moved, but Staples has never been underwater on a PPA contract.[10] While retail rates have not increased as much as they had earlier in the PPA contracts, Staples still reports a difference between PPA prices and retail rates in part because it contracted relatively low escalation rates (0.5%–2.0%). Staples receives additional benefits where it pays time-of-use rates, and, because its energy load profile matches well with its PV generation profile, it enjoys some demand-charge savings as well.[11] The company has 650 locations participating in demand-response programs. Although it is not actively controlling load in conjunction with PV installations at these facilities (e.g., turning off a building's air-conditioning compressor when clouds pass over and solar output drops), there is still a natural fit between demand-response and solar performance.

Today PV represents a very small percentage of Staples' total energy use. Still, although the company has explored other energy technologies, PV remains the company's focus for renewable energy in the U.S.

## 4. Economics of Financing PV: PPA Versus Self-Financing

In evaluating the decision to install PV, IKEA and Staples performed various economic analyses to determine the economic attractiveness of each proposal. The inputs for these analyses are private, and each project has had unique circumstances. In addition, data for these companies' PV systems and those of similar businesses are not uniformly reported. It is difficult, therefore, to analyze the specific economics of Staples' or IKEA's PV portfolio. However, modeling the economic attractiveness of PV ownership versus PPA contracting based on specific circumstances for a typical U.S. company illuminates the variables that might influence a company's decision. This section describes such an economic analysis. It also discusses qualitative factors that might influence a business. The analysis results do not suggest an option that is most efficient for every circumstance; rather, it highlights key factors a company or individual might consider in choosing a financing method.

## 4.1. Pro-forma Financial Models

### *4.1.1. Model Descriptions*

To compare the economics of financing commercial PV systems through self-financing versus a typical PPA, we built two pro-forma financial models: 1) a model of a commercial PPA and 2) a model of a commercial entity purchasing the PV system directly. Each model solves for the levelized cost of energy (LCOE) that satisfies all assumptions outlined in Table 1. For a more detailed set of assumptions, see Appendix A. In effect, LCOE provides comparisons between the models, acting as a proxy for the local utility retail rates necessary to satisfy the return requirements of investors and customers.[12] Annual payments during the term of the contract were also calculated to estimate how each scenario affects cash flow.

**Table 1. Basic Model Assumptions**

| Characteristic | Value |
|---|---|
| Installed PV price | $2.39/WDC |
| Location | San Diego, CA |
| Size | 500 kW |
| Project lifetime (PPA length) | 20 years |
| Incentives | 30% federal tax credit, 5-year Modified Accelerated Cost Recovery System (MACRS) depreciation schedule; no state incentives, including REC sales |
| Third-party capital cost (after-tax) | 9.2% return (9.0% tax equity, 10.5% sponsor equity) |
| Host business capital cost | 10% pre-tax return (6% after-tax) with sensitivity cases of 15% and 20% pre-tax returns[13] |

All models assume system location in California. Not only is California the largest U.S. solar market, but it is also economically feasible to install systems there without state incentives, allowing for the exclusion of this complicating factor from the models. Systems were assumed to have a 20-year economic life to align with the typical term length of a PPA. The models also calculate cash flows in years 20 through 30 to measure additional benefits beyond the life of the contract but within what is typically considered the lifetime of a PV asset.

The discount rates for commercial customers were assumed to be 10% pre-tax and 6% after-tax based on assumptions and calculations outlined in Appendix A. For a sensitivity analysis, the models used pre-tax discount rates of 15% and 23% as well. Although third-party providers might have additional costs (mostly attributable to the cost of capital associated with deploying TPO systems), we assume that system price and operating expenses are not affected by differences in financing method.[14]

### *4.1.2. Commercial PPA Pro-forma Financial Model*

In the PPA pro-forma model, the contract price per kilowatt-hour given to commercial customers was determined by the calculated LCOE reduced by the customer's discount rate, which is increased by the commercial entity's tax rate so as to provide it with an after-tax return comparable with self-ownership using a loan: PPA = LCOE × [1 – discount rate / (1 – tax rate)].

It was assumed that projects were financed using a sale-leaseback transaction between the developer (sponsor) and a tax equity provider.[15] The sponsor funds 15% of the project cost (in the form of an upfront lease prepayment which the sponsor expenses over the contract lifetime).

The sponsor is required to make yearly lease payments to the tax equity provider and to pay operating expenses associated with the project. In exchange, it receives all PPA revenues from the system host. Based on industry data, the sponsor is assumed to require a 10.5% after-tax rate of return. The tax equity provider funds 85% of the asset, net of lease prepayments, and in exchange, receives lease payments from the developer, an investment tax credit, and 5-year MACRS depreciation associated with the project. Based on industry data, the tax equity provider is assumed to require a 9.0% after-tax rate of return (Martin 2014). In the model, the developer leases the system for 80% of the contract term, at the end of which it purchases the system from the tax equity investor for 20% of the original installed price. This ensures that the arrangement qualifies as an "operating lease" rather than "capital lease."[16] The net cash flow providing the return to each participant is calculated on an after-tax basis.

### *4.1.3. Commercial Self-financing Pro-forma Financial Model*

In the commercial self-financing pro-forma model, the business funds the full cost of the project in year zero, but receives a 30% investment tax credit at that time as well. The business receives no revenue from the PV system, so its yearly benefit is calculated as electricity expense savings, or the calculated LCOE (i.e., comparable utility rate) multiplied by the expected PV system production. The LCOE is adjusted to provide the business with a 6% after-tax return. The business is responsible for all operating expenses, including O&M. It also expenses the cost basis of the PV system using the 5-year MACRS schedule. However, because utility expenses reduce the operating income of a commercial entity, the utility bill savings from PV produce higher operating income and thus higher taxes. The model captures this impact as well.[17]

## 4.2. Pro-forma Model Results

Commercial retail electricity rates vary dramatically depending on customer load profile; thus we did not compare the calculated LCOE for a PPA or self-financed system to a specific commercial retail rate. However, based on the assumptions outlined in Table 1 and Appendix A, the LCOE for the self-financed system was approximately 30% lower than the LCOE for the PPA-financed system owing to the higher cost of capital necessary for the sponsor and tax equity provider in a PPA transaction (Figure 6).

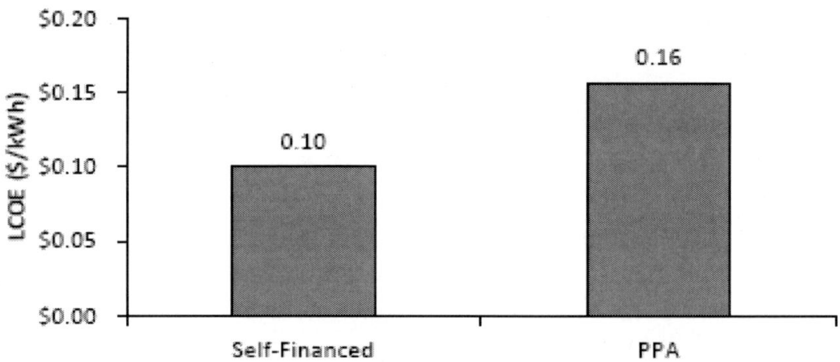


Figure 8. Modeled LCOE of commercial PV systems: self-financed versus PPA.

The results in Figure 6 are based on specific assumptions, and changing any of these affects the results. For example, under both financing methods, our base case assumes that the commercial customer has a pre-tax cost of capital of 10% (6% after-tax), which is based on analysis performed for a typical company. However, some companies have higher or lower costs of capital. In addition, a company might view a PV system as a riskier investment and thus require a higher rate of return while other companies might view PV as a safer investment, requiring a lower rate of return. When the commercial customer's pre-tax discount rate rises to 23%, the cost of energy is equivalent under either financing method (Figure 7). In other words, as the perceived risk of the project increases, the required rate of return increases, so future cash flows are discounted, requiring a higher LCOE in order to justify the upfront cash expenditure.

Another factor is the perceived risk of the financing method itself. If a commercial customer perceives system ownership to be riskier than TPO (as Staples did—see Section 3), then the different financing methods would have different discount rates. If, for example, a company assumed a 10% pre-tax discount rate for TPO and a 23% pre-tax discount rate for self-financing, then the LCOE would be 14% lower using a PPA (Figure 7).

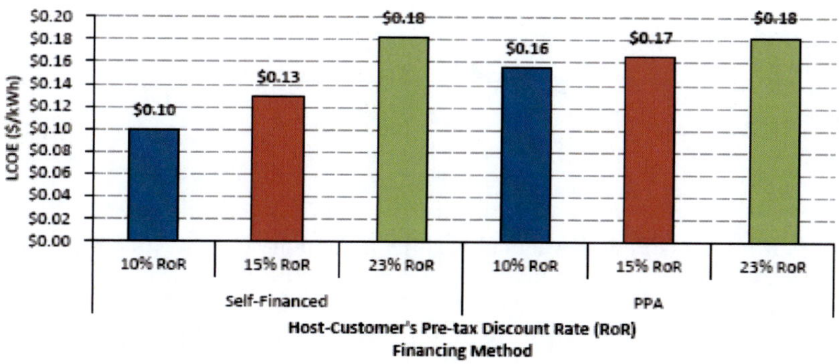


Figure 9. Modeled LCOE of commercial PV systems: self-financed versus PPA, with varying discount rates.

The timeline for customer payback is another consideration. A self-financed system is net-cashflow negative for approximately its first 5-11 years, depending on a company's hurdle rate, and is economically less attractive than a PPA system until year 6-14 (Figure 8).[18] Companies, such as IKEA (see Section 2), that view PV systems as long-term investments might be more likely than companies with a shorter-term planning horizon to accept a long period of negative cash flow in return for a better lifetime return.

## 4.3. Economic Factors Beyond the Pro-forma Financial Models

The pro-forma models demonstrate that, at commercial host-customer costs of capital below 23% (pre-tax), self-financing provides more attractive returns than a PPA over the long term.

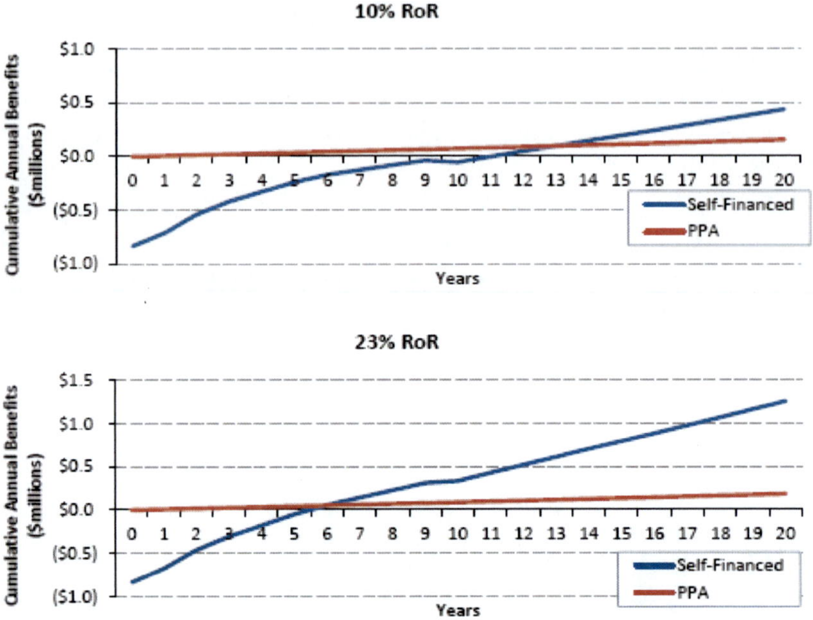


Figure 10. Modeled cumulative annual benefits for commercial PV systems: self-financed versus PPA.

This result, however, depends on the assumed rates of return for the third-party financiers and the host-customer, which might vary for several reasons. For example, PPA providers are trying to lower their cost of capital through innovative financing mechanisms, such as securitizing project portfolios. SolarCity, a residential and commercial PPA provider, recently raised funds from three securitizations at coupon rates of 4.0%–4.8%, backed by cash flows from pools of its distributed PV systems—capital that could fund future PV systems. Although the 4.0%–4.8% rates are not exactly comparable to the required returns in the model, as tax equity investors are still more than likely part of the transaction, they represent a significant reduction in the cost of capital normally required for TPO financing of a portfolio of PV assets.

Additional factors might make self-financing more attractive than a PPA. Our modeled PV systems have an economic life of 20 years while the average PV system is expected to last for at least 30 years.[19] A PV host that purchases a system owns it from the beginning. At the end of a PPA contract, however, the host must purchase the system, sign a new PPA, or allow the system to be removed from its property. We estimated the additional value of ownership in years 20-30 by comparing the modeled self-financed LCOE after 20 years and after 30 years, and we found that accounting for the full PV life lowers the LCOE by 17%. However, this is dependent on a company remaining in a building for 30 years (or being able to sell the system at some point in the future to the next building occupant/owner based on the remaining value of the system's expected 30-year life). If a company has a shorter time horizon (and/or believes there will be no resale value), a lease may be more attractive.

Although our models assume that system costs remain the same regardless of financing structure, self-financed systems typically involve fewer parties in the transaction than do PPAs and thus might be able to build and finance a system at a lower price. In addition, TPO systems often require the use of industrial-grade electric meters, which may add costs relative to the meters needed for a host-owned system. Feldman et al. (2013) estimate that TPO-related costs add $0.67/W to a commercial portfolio. This is often particularly true for small- to mid-sized commercial systems (100kW-500kW) not part of a larger portfolio. These systems are often more complicated than residential-sized systems (5kW-10kW), but do not provide economies of scale, making them very expensive for TPO.

However, there are also many reasons a company may find a PPA more beneficial than self-ownership. While a system owner may have a lower cost of capital, the high upfront cost and long payback period may be unattractive. This is particularly true for agents with time horizons shorter than the system's lifetime. For example, a company's facilities manager, who is unlikely to remain in the same job for the economic life of the PV, might choose immediate energy savings over potential long-term economic value. A company might also require immediate savings to approve a project.

Commercial entities must also consider how various financing options impact their financial statements. While PPAs and system purchases are both long-term commitments, PPAs are considered off-balance sheet transactions; similar to an operating lease, PPA payments are treated as operating expenses, and the long-term liability of the contract does not appear on a company's balance sheet.[20] Financial statements are used to measure the financial health of a company by internal and external parties, and corporate debt often has covenants that limit the amount of additional debt a company can borrow. Therefore, a company might be unable to add the liability of a PV loan onto its balance sheet and might instead opt for a PPA. Much of this depends on the size of the company relative to the amount of PV it aims to deploy.

In addition, a company that self-finances must have sufficient tax liability to take advantage of the 30% federal tax credit and 5-year MACRS depreciation schedule, which could be worth up to an additional 25%–30%, depending on the company's tax rate. Although the benefits can be carried forward for use in later years, their future worth is lower owing to the time-value of money.[21] Businesses also pay varying amounts of federal taxes, although the degree to which an individual business can use a credit largely depends on business size, PV system size, and the business's ability to adjust incurred income.

Businesses that own PV and those that finance it through a PPA also face different risk and reward considerations. To a business with a PPA, the only risk to unexpectedly low PV generation is the cost of additional electricity that must be sourced from the utility. In contrast, a PV owner must pay for the system, including O&M, regardless of system performance (unless there is a production guarantee). If state and/or local incentives are required to make a system economical, system owners may also be exposed to issues such as Renewable Energy Certificate (REC) price volatility and delays in disbursements from grant programs. Further, there is a risk of interconnection delays and lost production while waiting for the utility to approve the system's operation.[22]

A system owner might also incur unforeseen O&M costs. For example, inverters do not typically have warranties for the full life of a PV system (which is why our model assumes an inverter replacement in year 10). While a system with a PPA may similarly require an inverter replacement in year 10, this cost is built into the PPA price. A system owner, on the other hand, must pay all O&M costs and so, in any given year, might incur costs higher than would have been incurred using a PPA.

Many of these risks of PV ownership are small (such as system production), can be hedged against (for example, by selling long-term REC contracts[23]), or can be borne by another party (for example, by requiring a production guarantee from the installer or delaying payment for the system until receipt of state rebates and interconnection). As the PV industry matures, and more historical data is made available, there may be a reduction in the overall perception of risk.[24] Regardless, customers likely must educate themselves about PV risks and benefits more when they are buying a system than when they are entering into a PPA or other TPO contract.

## CONCLUSION

Although Staples and IKEA are both among the largest PV system hosts in the United States, the companies have used different approaches to PV financing. IKEA owns all of its PV systems, a decision that fits well with its corporate culture and business model. This ownership model spurred the company to engage in the entire PV deployment process, from system feasibility through O&M, identifying opportunities and risks at each stage. Over time, its process evolved, and its personnel gained valuable knowledge and skill. IKEA assumed more risk by owning PV instead of leasing it (for example, if the company had an inability to use the tax credits generated). However, through careful management of its processes, IKEA believes it is receiving the full benefits of owning long-term, high-quality assets.

In contrast, Staples leases all of its PV systems through PPAs, which aligns well with the company's risk-return preferences.[25] Still, Staples has actively managed its PV deployment, developing the skills to analyze the economics and feasibility of PV in relation to its buildings, local electricity rates, government incentives, and the PV market in various locations. It also has benefited from the technical and market expertise of the third-party owner, which has enabled more rapid PV deployment. A high percentage of leased properties and the potential for future site relocation are hindering PV development in some Staples locations. The company, however, is pursuing several strategies for increasing its share of PV-generated electricity, including virtual net-metering, deployment of large systems at distribution centers, solar-friendly leases on new property, and collaboration with building owners who provide PPAs coterminous with their leases.

Our modeling quantifies some of the financial drivers that companies like IKEA and Staples consider when choosing a PV financing strategy. Based on specific assumptions—including a 10% pre-tax cost of capital for the commercial customer—the LCOE for the self-financed system is approximately 30% lower than the LCOE for the PPA-financed system. However, the attractiveness of the financing methods varies due to different assumptions for cost of capital, different perceptions of PV investment risk, and different perceptions of the risk of using a particular financing method. When the commercial customer's pre-tax discount rate rises to 23%, for example, the cost of energy is equivalent under either financing method. If a company assumed a 10% pre-tax discount rate for TPO and a 23% pre-tax discount rate for self-financing, then the LCOE would be 14% lower using a PPA.

The timeline for customer payback is another consideration. Although self-financing provides a better return over the 20-year project lifetime in our model, it is net-cash-flow negative for about 5-11 years. The PPA, on the other hand, produces positive cash flow quickly and is economically more attractive than the self-financed system for the first 6-14 years. Still, our modeled self-financing is even more attractive over the long term if the PV lifetime is extended to an expected period of 30 years—this assumption lowers the LCOE by 17%. Self-financing might also have advantages over PPAs due to involving fewer parties in the financial transaction and requiring less-expensive electric meters.

That said, PPAs offer several potential advantages to some companies. Unlike purchase of a PV system, PPAs are considered off-balance sheet transactions, which might be attractive for companies unwilling or unable to add the liability of a PV loan onto their balance sheets. In addition, a company can get the full benefits of a PPA with any amount of tax liability whereas self-financing companies must have sufficiently large tax liability to take full advantage of PV tax incentives. Companies using PPAs also might reduce their exposure to risks related to PV underperformance, unanticipated O&M costs, and delays in receiving incentives and grid-interconnection approval.

Our analysis suggests that the most appropriate PV financing option for a particular business depends on the characteristics and circumstances of that business. Our modeling results and discussion, as well as the case studies of the IKEA and Staples PV experiences, might provide useful insights to other companies that are considering launching or expanding onsite PV programs.

# APPENDIX A

**Table A-1. Basic Assumptions for Commercial Pro-forma Financial Models**

| | |
|---|---|
| Installed Price | \$2.39/WDC [a] |
| Capacity Factor | 17.3% [b] |
| Inverter Replacement | \$0.13/W [c] |
| Combined Federal & State Tax Rate | 40.2% |
| Annual Degradation Rate | 0.50% [d] |
| Project Lifetime (PPA length) | 20 years |
| O&M (% of installed cost) | 0.50% [e] |
| Inflation | 2% |
| Federal Tax Incentives | 30% Investment Tax Credit & MACRS Depreciation Schedule |

[a] Source: SEIA (2014b).

[b] Capacity factor calculated by PVWatts (version 1) for a PV system in San Diego, with a derate of 0.85, azimuth of 180 degrees, and tilt of 5 degrees.

[c] Assumes that, when inverter replacement occurs in year 10 of the project, prices have dropped to targets outlined in The SunShot Vision Study, adjusted for inflation (DOE 2012).

[d] Source: Bolinger (2014).

[e] Source: Shah et al. (2013).

**Table A-2. Assumptions Specific to PPA Pro-forma Financial Models Using an All-equity Sale-leaseback Transaction for Commercial Projects**

| | Tax Equity | Sponsor Equity |
|---|---|---|
| Ownership Percentage | 85% [a] | 15% |
| After-tax Rate of Return | 9.0% [b] | 10.5% [c] |
| Lease Length | 16 years [d] | |
| Residual Value | 20%e | |

[a] Source: Bolinger (2014).

[b] Source: Martin (2014).

[c] After-tax cost of capital for SolarCity as calculated by the Capital Asset Pricing Model (CAPM); see below for calculations.

[d] Source: Peterson (2012).

**Table A-3. Assumptions for Commercial Customer's Cost of Capital Using Capital Asset Pricing Model (CAPM)**

| Variable | Value | Explanation |
|---|---|---|
| Risk-free Rate (1) | 3.6% | 20-Year Treasury Bill Constant Maturity. December 1, 2013.<br>Source: FRED 2014. |
| Equity Beta (2) | 1.00 | Assumes an equity beta equal to the market as a whole. |
| Debt/Asset Ratio (3) | 50% | Source: Newell et al. 2012. |
| Tax Rate (4) | 40.2% | Assumes a federal corporate income tax rate of 35% and state corporate income tax rate of 8%. |
| Equity Risk Premium (5) | 5.3% | Source: Credit Suisse 2013. |
| After-tax Cost of Equity (6) | 8.9% | = (1) + (2)*(5) |
| Debt Premium (7) | 1.4% | = (8) - (1) |
| Pre-tax Cost of Debt (8) | 5.1% | Annual Moody's Seasoned Aaa Corporate Bond Yield, 2004- 2013. Source: FRED 2014. |
| Nominal After-tax Weighted Cost of Capital (9) | 6.0% | = (1-(3))*(6)+(3)*(8)*(1-(4)) |
| Nominal Pre-tax Weighted Cost of Capital (10) | 10.0% | = (6) / (1-(4))*(1-(3))+(8)*(3) |

# References


Bolinger, M. (2014). An Analysis of the Costs, Benefits, and Implications of Different Approaches to Capturing the Value of Renewable Energy Tax Incentives. Berkeley, CA: Lawrence Berkeley National Laboratory.

Buckley, M. (Aug. 14, 2014). Telephone conversation with Feldman, D. and Margolis, M., *National Renewable Energy Laboratory.* Staples Inc., Framingham, MA.

Clean Energy Finance and Investment Authority (CEFIA). (undated). "Staples, Inc." Accessed October 2014: http://www.ctcleanenergy.com/Your BusinessorInstitution/CommercialInstallations/DistributionFacilityInstallations/StaplesKillingly/tabid/442/Default.aspx.

Credit Suisse Research Institute. (2013). "Credit Suisse Global Investment Returns Yearbook 2013." Accessed November 2014: http://www.investmenteurope.net/wpcontent/uploads/2014/07/2013_yearbook_final_web.pdf.

CSI (California Solar Initiative). (2014). Current CSI Data. Accessed April 4, 2014: www.californiasolarstatistics.ca.gov/current data files/.

DOE (U.S. Department of Energy). (2012). SunShot Vision Study. DOE Report No. DOE/GO102010-3037. Washington, DC: U.S. Department of Energy.

EIA (U.S. Energy Information Administration). (2014). Form EIA-861, Annual Electric Power Industry Report; and Form EIA-861S, Annual Electric Power Industry Report (Short Form). Table 5.1. Retail Sales of Electricity to Ultimate Customers. Accessed September 3, 2014. http://www.eia.gov/electricity/data.cfm#sales

Feldman, D.; Friedman, B.; Margolis, R. (2013). Financing, Overhead, and Profit: An In-Depth Discussion of Costs Associated with Third-Party Financing of Residential and Commercial Photovoltaic Systems. NREL Report No. NREL/TP-6A20-60401. Golden, CO: *National Renewable Energy Laboratory.*

Feldman, D.; Lowder, T. (2014). Banking on Solar: An Analysis of Banking Opportunities in the U.S. *Distributed Photovoltaic Market. NREL/TP-6A20-62605. Golden, CO: National Renewable Energy Laboratory.*

"Federal Reserve Economic Data (FRED)." (2014). Economic Research Division. Federal Reserve Bank of St. Louis. Accessed November 2014: http://research.stlouisfed.org/fred2.

Gasper, M.; Roth, J. (May 15, 2014). Telephone conversation with Feldman, D. and Margolis, M., National Renewable Energy Laboratory. IKEA Property, Inc., Conshohocken, PA.

Gasper, M. (Nov. 17, 2014). Email to Feldman, D. National Renewable Energy Laboratory. Ikea Property, Inc., Conshohocken, PA.

IKEA. (2014). "IKEA US Renewable Energy Commitment Included In President Obama Announcement of Commitments and Executive Actions to Advance Solar Deployment and Energy Efficiency." IKEA Corporate News, May 8, 2014: http://www.ikea.com/us/en/about ikea/newsitem/ President Obama renewable energy.

Martin, K. (2014). "Cost of Capital: 2014 Outlook." Chadbourne & Parke Project Finance News, February 2014.

Massachusetts Department of Energy Resources (DOER). (2014). "RPS Solar Carve-Out Qualified Units." Accessed March 26, 2014: http://www.mass.gov/eea/docs/doer/rps-aps/solarcarve-out-units.xlsx.

Newell, S.; Spees, K.; Pfeifenberger, J.; Mudge, R.; DeLucia, M.; Carlton, R. (2012). ERCOT Investment Incentives and Resource Adequacy. The Brattle Group.

NJCEP (New Jersey Clean Energy Program). (2014). NJCEP Project List as of March 31, 2014. Accessed July 9, 2014: http://www.njcleanenergy.com/renewable-energy/home/home.

Ong, S.; Denholm, P.; Doris, E. (2010). Impacts of Commercial Electric Utility Rate Structure Elements on the Economics of Photovoltaic Systems. NREL Report No. TP-6A2-46782. Golden, CO: *National Renewable Energy Laboratory.*

Peterson, L. (2012). "Primary Tax Equity Finance Structures Common to the U.S. Domestic Solar Energy Industry: June 2012." Reznick Group.

PJM Environmental Information Services (PJM EIS). (2014). "Generation Attribute Tracking System (GATS): Renewable Generators Registered in GATS." Accessed November 12, 2014: https://gats.pjm-eis.com/gats2/PublicReports/RenewableGeneratorsRegisteredinGATS.

SEIA and GTM Research (Solar Energy Industries Association and GTM Research). (2014). U.S. Solar Market Insight Report: 2013 Year-In-Review. Washington, D.C.: Solar Energy Industries Association.

SEIA and GTM Research. (2014b). U.S. Solar Market Insight Report: Q2 2014. Washington, D.C.: Solar Energy Industries Association.

SEIA and Vote Solar. (Solar Energy Industries Association and The Vote Solar Initiative). (2013). Solar Means Business 2013: Top Commercial Solar Users. Washington, D.C.: Solar Energy Industries Association.

StrataSolar. (2014). "IKEA – 1 MW – ATLANTA." Accessed October 2014: http://www.stratasolar.com/spotlight/commercial-retail-atlanta-ga/.

Shah, V.; Booream-Phelps, J.; Min, S. (2013). "Distributed Generation to Herald New US Growth Era." Deutsche Bank Market Research.

Wiser, R..H.; Mills, A.D.; Barbose, G.L.; Golove, W.H. (2007).The Impact of Retail Rate Structures on the Economics of Commercial Photovoltaic Systems in California. LBNL-63019. Berkeley, CA: *Lawrence Berkeley National Laboratory.*

## End Notes

[1] In this paper we use the term "lease" to refer to both 1) customers that pay a set fee per month for the use of an onsite PV system; and 2) PPA customers that pay per kWh of generated electricity from an onsite PV system. In some jurisdictions there is a legal distinction between the two arrangements; however, for the purposes of this paper, we are using the term "lease" in this broader sense.

[2] See Appendix A for weighted Average Cost of Capital (WACC) calculations.

[3] This figure includes multiple floors within buildings but excludes land.

[4] This is in the case of self-ownership. TPO transactions, in contrast, are off the balance sheet.

[5] IKEA does have franchises, but it franchises to itself.

[6] The exceptions to this are the four distribution centers with solar that, due to building size and thus rooftop area, hold very large arrays while also consuming much less energy than IKEA stores do. A big distribution center, unlike a typical store, is mostly a non-air-conditioned warehouse space; stores have display lights, full air conditioning, restaurant components, vertical transport, etc.

[7] Net-metering rules vary significantly across the country, sometimes involving fixed fees, capacity limits, rules on rolling credits forward, and other nuances that can impact the economics of installing a system on a commercial building. Additionally, utility rate structures also vary across the country, sometimes involving fixed fees, capacity fees, or energy fees. PV system economics can change dramatically depending on the applicable net-metering rules and utility rate structures, as well as the load profiles of the building and the PV system. For a detailed discussion of the potential impacts of net metering and utility rate structures on PV system economics see Ong et al. (2010) and Wiser et al. (2007).

[8] Federal tax incentives significantly limit the sale of solar assets in their first 5 years.

[9] The company missed early opportunities in places like Germany, says Buckley (Buckley, 2014).

[10] "Underwater" refers to when the PPA price is higher than current retail electricity rates.

[11] When negotiating its PPAs, Staples expected there would be demand savings but did not know how much, so this was not factored into the original economic analysis or PPA contracts.

[12] In other words, an LCOE is calculated to give each party (installer, customer, third-party owner) the required return necessary for them to participate in the transaction. These calculated LCOE's represent the minimum retail rate at which these solar projects are economically viable (given the assumptions shown in Table 1); and therefore, the lower the

minimum retail rate (LCOE), the larger the potential market (or the more savings a customer can expect).

[13] The assumptions are not based on specific project details from IKEA or Staples; however, the self-financed scenario is similar to IKEA's financing model in that it does not use any project-level debt to finance the PV system. Instead, debt was assumed at the corporate level, as shown in the WACC calculations in Appendix A.

[14] Self-owned and TPO projects may also differ in other ways, such as different O&M costs, tax rates, as well as the assumed basis for which the tax credit and depreciation are calculated.

[15] There are currently several popular structures for financing a tax equity investment in a solar asset: sale-leaseback, partnership flip, and inverted lease. The choice between these structures is in large part affected by the preferences of the tax equity provider and the needs and ability of the developer or sponsor. The sale-leaseback structure was chosen for this analysis because of its prevalence in the marketplace; however, a different financing structure would not affect the general outcome of this analysis assuming the same financial hurdle rates apply.

[16] In capital leases, the lessee effectively owns the asset under tax and accounting rules.

[17] In reality, the reduction in electricity expenses may not result in the full loss of the expense multiplied by the corporate tax rate due to the complexity with which companies pay their taxes.

[18] Note that companies can get a loan for a PV system, avoiding the upfront negative cash flows. A more detailed discussion of solar loans can be found in Banking on Solar: An Analysis of Banking Opportunities in the U.S. Distributed Photovoltaic Market (Feldman & Lowder 2014). Also, the cash flows in the "23% RoR" case of Figure 8 are higher than those in the "10% RoR" case. This is because a higher return requires more cash to achieve this higher rate of return, which is why the required LCOEs are greater as well.

[19] As a point of reference, most PV module warranties have 25-year terms.

[20] There has been an effort within the Financial Accounting Standards Board (FASB), which established Generally Accepted Accounting Principles (GAAP), to refine lease accounting standards to remove the distinction between operating and capital leases--in effect, moving all leases on balance sheet. However, this effort has been underway for a while and no final decisions have been made.

[21] The business tax credit can be carried forward 20 years.

[22] There is also the risk of changing rate structures, such as changes in net-metering provisions, but this risk is mostly shared by system owners and PPA holders.

[23] It should be noted that in many markets it is currently hard to find a counterparty for long-term REC contracts.

[24] NREL and DOE are working on accelerating this process through several initiatives, by making more data available and creating better transparency through improved documentation and procedures. These include, among others, the Solar Access to Public Capital (SAPC) working group. For more information, please see: https://financere.nrel.gov/finance/solar_securitization_public_capital_finance.

[25] Staples also began installing systems earlier than IKEA, when PV systems were more expensive, which may have impacted their decision to finance PV through a PPA.

In: To Own or Lease Solar Installations
Editor: Jordan Gomez

ISBN: 978-1-63483-080-5


*Chapter 2*

# BANKING ON SOLAR: AN ANALYSIS OF BANKING OPPORTUNITIES IN THE U.S. DISTRIBUTED PHOTOVOLTAIC MARKET*

***David Feldman and Travis Lowder***
National Renewable Energy Laboratory

## LIST OF ACRONYMS

| | |
|---|---|
| ABA | American Banker's Association |
| APR | annual percentage rate |
| CEFIA | Clean Energy Finance and Investment Authority |
| DOE | U.S. Department of Energy |
| FDIC | Federal Deposit Insurance Corporation |
| Fed | United States Federal Reserve |
| GDP | gross domestic product |
| HELOC | home equity line of credit |
| ITC | investment tax credit |
| LCOE | levelized cost of energy |
| MACRS | modified accelerated cost recovery system |
| NCUA | National Credit Union Administration |
| O&M | operations and maintenance |

* This is an edited, reformatted and augmented version of technical report NREL/TP-6A20-62605, issued by the National Renewable Energy Laboratory, November 2014.

| | |
|---|---|
| OCC | Office of the Comptroller of the Currency |
| PACE | property assessed clean energy |
| PBI | production-based incentive |
| P&I | principal and interest |
| PPA | power purchase agreement |
| PV | photovoltaic |
| REC | renewable energy credit |
| RPS | renewable portfolio standard |
| TPO | third-party ownership |
| UCC | Uniform Commercial Code |
| WACC | weighted average cost of capital |

# EXECUTIVE SUMMARY

This report provides a high-level overview of the developing U.S. solar loan product landscape, from both a market and economic perspective. It covers current and potential U.S. solar lending institutions; currently available loan products; loan program structures and post-loan origination options; risks and uncertainties of the solar asset class as it pertains to lenders; and an economic analysis comparing loan products to third party-financed systems in California.

Solar-specific loan financing is growing in the United States—2014 in particular saw several new loan product announcements and program launches (See Footnote 2 in the Introduction). A solar loan financing arrangement differs from third-party ownership (TPO) in several key aspects, including: the retention of ownership rights by the system host and its associated tax benefits and other incentives; the fixed nature of its monthly payment (similar to a lease but not a power purchase agreement [PPA]); and the variability in the size of payments based on the interest rate and tenor of the loan (i.e., individual payments spread over a longer period will be smaller in size).

Several analysts and industry stakeholders have indicated that solar loans will increasingly capture market share relative to the TPO model in the coming years. While the actual competitiveness of the loan option in solar finance will be determined by the offerings in the market, this report attempts to provide a framework for understanding how loan structures could affect the ultimate cost to distributed PV consumers. Solar loans have the potential to provide an economical option (from an LCOE perspective) for homeowners and businesses to finance the purchase of a solar system, retaining the benefits of ownership that TPO systems cannot provide while avoiding the large upfront cost of a PV system.

Section 4 of this report presents the results of an economic analysis comparing the economics of residential and commercial customers using solar loans to those using TPO to finance on-site PV generation. As demonstrated in Figure ES-1 below, the levelized cost of energy (LCOE) for residential systems with solar loans was lower than the LCOE for residential systems with PPAs by 19% to 29% (varying by the term of the loan), due to the higher cost of capital necessary for the sponsor and tax equity in a PPA transaction.

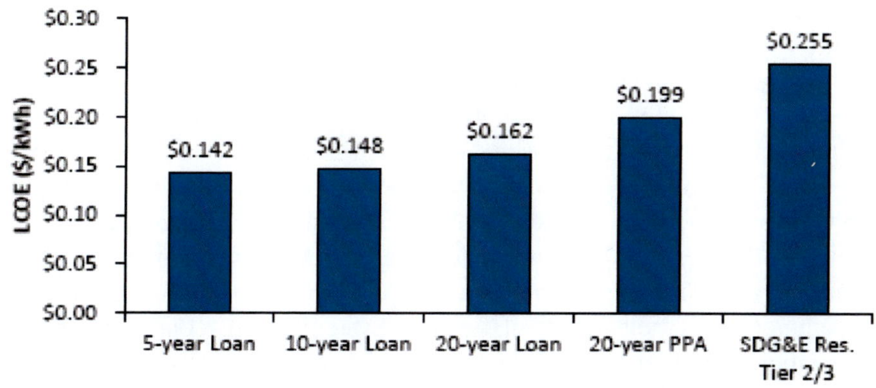


Figure ES-1. LCOE of residential PV systems, financed under a PPA or loan, compared to retail rates.

There are, however, several economic factors which influence the value of loans to consumers that go beyond the pro forma financial models. These include the higher annual payments to service these shorter-term loans during the beginning of the lifetime of the solar asset – as demonstrated in Figure ES-2 below. In fact, in the first year of a five-year loan, the debt service payments were calculated to be almost double what a customer would pay to a utility, and over twice as much as a PPA or 20-year loan. At the end of the loan term, a customer's annual payments would drop to the cost of operations and maintenance (O&M), including an inverter replacement in year 10, and, if desired, monitoring by a third party; however, some customers may not want to pay more for their electricity for the first five or ten years of the PV system's operation.

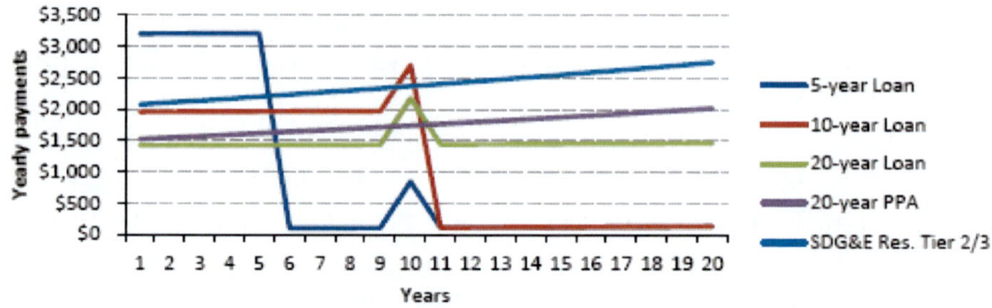


Figure ES-2. Comparison of annual payment between residential PV systems, financed under a PPA or loan, and retail rates.

Other factors that may impact the economics of a solar loan or TPO include whether or not the host has sufficient tax liability to take advantage of the federal and/or state tax incentives; the additional liability and maintenance costs associated with ownership; the complications of adding assets and liabilities onto one's balance sheet (as opposed to a TPO transaction, which is off-balance sheet); and the economic time horizon of individual decision makers. Moreover, any specific market offerings in either solar loans or TPO products will differ from the assumptions in this report, which provides only a general framework for interpreting the comparison and relative competitiveness of both forms of finance. System costs, monthly payments, amortization schedules, the ultimate cost of energy, and other factors will differ by each product and each individual deal with the consumer.

# 1. INTRODUCTION

The growth in distributed (and particularly residential) solar photovoltaics (PV) deployment in the U.S. has been facilitated in large part through the third-party ownership (TPO) model. In 2013, TPO represented some 66% of the U.S. residential solar market, and a considerable portion of the commercial market (Litvak 2014). The success of the TPO model is attributable in part to its economic proposition: TPO can provide consumers access to PV-generated electricity at a price that is competitive with those consumers' utility (i.e., retail) rates. However, in the last two years, another solar financing option is becoming commensurately competitive and has begun to capture market share: loans.

Until recently, loan financing for distributed solar installations was largely done through home equity loans or home equity lines of credit (HELOCs), commercial loans, and other standardized loan products available to homeowners and businesses for general expenditures. Historically, solar-specific loans—i.e., products for which the underwriting, loan terms, lender security interest, and other programmatic aspects are all designed for financing solar installations exclusively[1]—have not had much market presence. Prior to the fall of 2013, there were few widely available (i.e., not jurisdictionally limited, such as property assessed clean energy programs or on-bill financing in a utility's service territory) solar loan options in the United States. However, from the period between approximately October 2013 and October 2014, at least nine new solar-specific loan programs were announced, and several more have begun operations without a formal announcement.[2]

A solar loan financing arrangement differs from TPO in several key aspects, including:

- *Ownership:* When financing through a loan, the system host retains ownership of the PV assets. Third-party ownership, as the name implies, entails ownership and maintenance by the solar company and its investor partners.
- *Tax benefits and other incentives:* Owners of loan-financed systems receive all applicable tax benefits and incentives available to the solar assets (this follows from ownership). As of October 2014, these benefits and incentives could include: the 30% federal investment tax credit (ITC) for individuals (see Section 3), production-based incentives (PBIs), renewable energy credits (RECs), and others. When leasing a system through a TPO arrangement, these benefits typically go to the third party. Additionally, third-party owners can make use of the accelerated depreciation schedule to increase the system cost savings—homeowners cannot.
- *Monthly payments:* A solar loan is typically amortized through monthly payments of both principal and interest (P&I). In contrast, TPO systems are paid for via either a monthly fixed rate in a lease arrangement, or a charge per unit of electricity produced by the system (on a \$/kWh basis) in a PPA arrangement. It is common for both leases and PPAs to contain annual escalators that step up the charges by a specified percentage (typically around 3%) each year.
- *Effect on home valuation:* Some studies suggest that homeowners could increase the value of their home when they install and own a

solar system on their rooftop (Hoen et al. 2013; Desmarais 2013). While no comparable study has yet been performed on TPO systems, the National Renewable Energy Laboratory's (NREL's) communications with appraisers and homeowners in California, as well as speculations made in recent news articles (Wade 2014) suggest that systems financed through a PPA or a lease may not be rolled toward the value of the home (appraisers may, in fact, count them a deduction against the home value), and could complicate the sales process. It is important to note, however, that the effect that each financing option has on home value is still not well understood. Additional data, regulatory decision making, and the development of industry best practices will be necessary before the market arrives at a standard method for appraising solar assets.

- *Cost*: The cost to finance a system through a loan is largely determined by the interaction between the loan amount, the applicable interest rate, and the tenor or term of the loan. For example, a $20,000 loan with a 6% annual percentage rate (APR) of interest and a 15-year maturity will add an extra $18,000 in costs to the principal amount, or about an extra $100 a month in interest payments. TPO systems are typically financed on a portfolio basis through complex financial structures that bring in tax equity investors to monetize the 30% ITC and depreciation expense. The weighted average cost of capital (WACC) of these structures reflects, principally, the tax equity's yield on their investment, the cost of any associated debt, and the sponsor's (i.e., the solar company's) cost of equity. The resulting cost of the system will be influenced by this WACC, which is typically higher than the cost of capital on the loan.

This report provides a high-level overview of the developing U.S. solar loan product landscape, from both a market and economic perspective. It covers current and potential U.S. solar lending institutions; currently available loan products; loan program structures and post-loan origination options; risks and uncertainties of the solar asset class as it pertains to lenders; and an economic analysis comparing loan products to third party-financed systems in California. The report begins (Section 2) with an overview of the U.S. distributed solar market to contextualize the discussions to follow. For readers already familiar with this information, the authors recommend beginning at Section 3 on page 18.

The penultimate section of the report (Section 4) provides an economic analysis of solar loans versus TPO. The authors built three pro forma financial models to calculate the levelized cost of energy (LCOE) for a residential loan, a commercial loan, and a TPO PPA in the state of California. These LCOE figures are used to compare the system costs associated with each financing arrangement, and—for the residential loan and TPO PPA—to compare with the retail rates of one of California's three investor-owned utilities. This report finds that the single largest differentiating factor influencing LCOE in the models was the cost to finance the solar assets— namely the interest rate on the loan, and the WACC in the TPO PPA. Accordingly, the modeled LCOE for solar loans was lower than the TPO LCOE in this analysis.

# 2. The U.S. Solar Market

## 2.1. Solar Policy: Fundamentals for Growth

The relatively high levels of year-over-year growth recently achieved in the U.S. photovoltaic (PV) market is largely due to the financial incentives and support policies available from the federal, state and local governments. Historically, national incentives have been provided primarily through the U.S. tax code, in the form of a 30% ITC given in the first year of operation, either under section 48 or 25D of the tax code. The section 48 credit is used by businesses (e.g., all commercial and utility-scale installations and to third-party owned residential, government, or non-profit installations), and in 2017 will revert to 10%. The 25D credit is for persons using the solar property for residential purposes and, unlike the section 48 credit, will cease to exist in 2017. Businesses which claim the ITC have the additional benefit of using an accelerated 5-year tax depreciation schedule for the solar asset; together they are commonly referred to as "tax benefits."

At the state level, renewable portfolio standards (RPSs) have proven to be one of the most significant drivers of renewable energy deployment in the United States. An RPS, also called a renewable electricity standard (RES), requires electricity suppliers to purchase or generate a targeted amount of renewable energy by a certain date. Although design details can vary considerably, RPS policies typically enforce compliance through penalties, and many include the trading of renewable energy certificates (RECs). As of November 2013, seventeen states and Washington, D.C. had RPS policies with specific solar provisions (Barbose 2013).

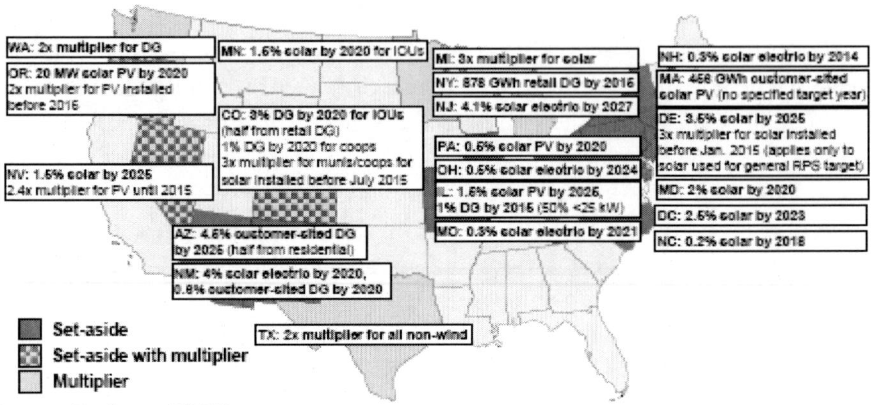


Source: Barbose (2013).
Note: Compliance years are designated by the calendar year in which they begin.

Figure 1. State renewable portfolio standards with solar and DG set-asides.

As an alternative to RECs, states have incentivized PV deployment through up-front cash grants, performance-based cash grants, state and local tax credits, and feed-in tariffs. Local jurisdictions without strong state solar mandates (e.g., Austin, Texas) have also developed solar initiatives as well. Recently, PV systems have been installed in certain markets in the U.S. (e.g., Hawaii, California) without the need for state or local incentives, either because retail electricity rates in those markets are relatively high, or the cost to install the systems has fallen to relatively low levels. As shown in Figure 2, in the fourth quarter of 2013, only 36% of distributed systems installed in California received assistance from the California Solar Initiative (CSI), the state's largest incentive program. This was down from 88% just two years earlier.

## 2.2. Third-party Ownership: A Market Solution

Given the current state of the United States PV market, benefiting from ownership of solar assets often requires, among other things:

- A large upfront investment of capital
- Sufficient expected taxable income to utilize a solar system's tax benefits

- Knowledge of the local and state permitting and incentive programs (due to the fractured nature of U.S. electric industry and PV incentive programs)
- An understanding of the potential risks and uncertainties associated with long-term ownership of a relatively new asset class.

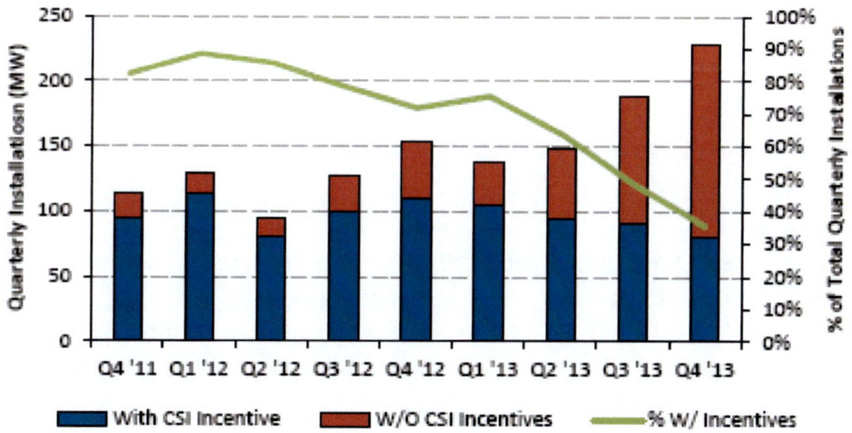


Sources: CSI (2014); SEIA & GTM (2014).

Figure 2. Distributed PV systems installed in California, with and without incentives, by quarter.

As state and local incentive programs wind down, or exhaust their budgets, many analysts expect a larger share of systems to be installed with only the federal incentives.

One business model that attempts to overcome these obstacles is third-party solar ownership. Under this arrangement, a third-party entity purchases, owns and operates the PV system on a homeowner or business's roof or property. In exchange, the homeowner or business signs a longterm contract (i.e., 15-25 years) to lease the system or purchase the electricity generated by the system (power purchase agreement, or PPA); typically at a rate less than the price of retail electricity rates. The homeowner or business benefits from on-site solar PV generation at or below electric utility costs, but without the upfront outlay of capital or any complications associated with operating a system. This model has gained significant market share in many states across the U.S., with third-party systems comprising approximately 60% to 80% of residential systems installed in California, Arizona, and Massachusetts – three of the top U.S. residential markets, as shown in Figure 3.[3]

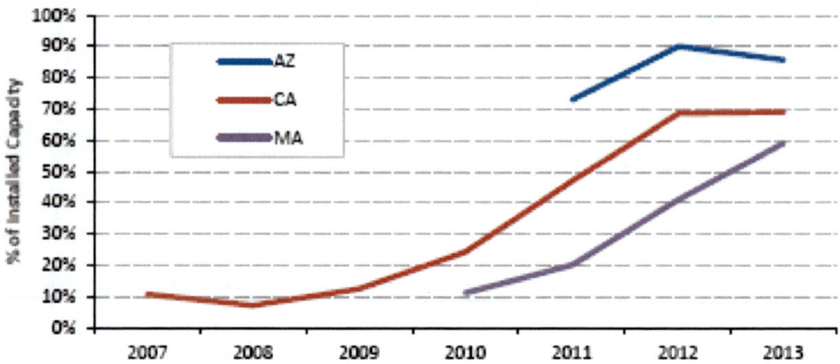


Sources: Arizona Goes Solar (2014); CSI (2014); DOER (2014).

Figure 3. Third-party market share of annual residential installations, by state.

During this time period the U.S. (and global) market has experienced a rapid reduction in PV system costs, across sectors. As shown in Figure 4, from 2010 to 2013 residential PV system price fell from around $6.5/W to below $4/W, a reduction of 42%. Price reductions in the commercial and utility-scale markets decreased by similar rates as well.

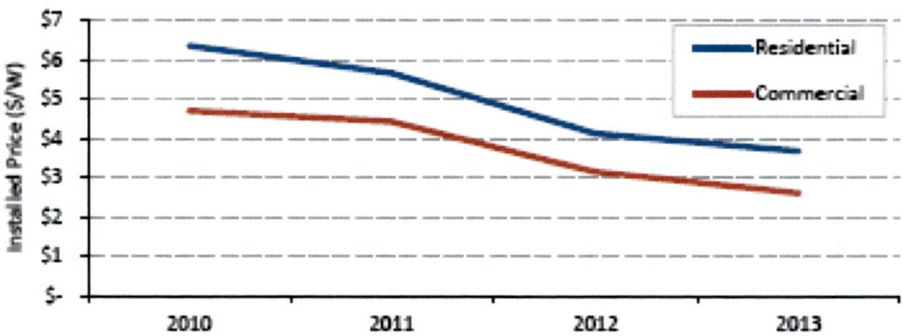


Source: Feldman et al. (2014a).

Figure 4. Average U.S. PV system price by market segment.

With innovations in financing, such as third-party ownership, and reductions in average system price, the U.S. PV market has grown dramatically. As shown in Figure 5, annual distributed U.S. PV installations have increased 13-fold in the past six years, from 150 MW in 2007 to 1,900 MW in 2013. Many analysts expect this growth to continue in the near future, with annual distributed installations potentially doubling or tripling by 2016.

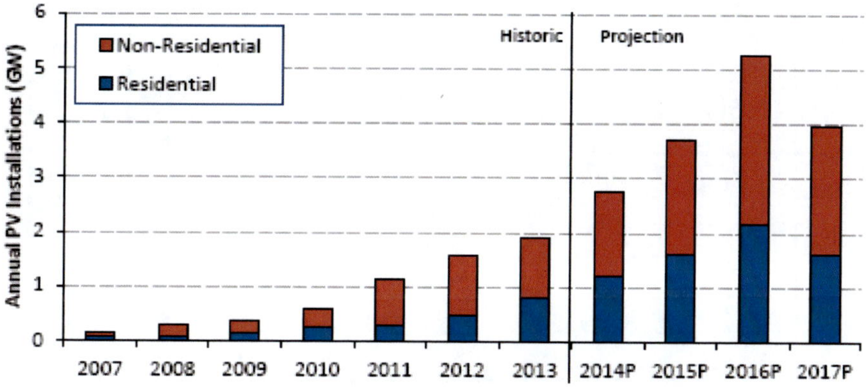


Sources: Lee et al. (2014); Linder et al. (2014); SEIA & GTM (2014).

Figure 5. Historic and projected annual distributed U.S. PV installations by sector: 2007-2016.

While analysts do expect a reduction in demand in 2017 when the 30% ITC expires (or is reduced to 10%), several market conditions support the idea that the U.S. PV industry will remain strong:

- PV industry corporate filings have noted that their business models are preparing for the ITC expiration and are working to reduce costs to remain competitive in a post-incentive business environment.
- The U.S. currently has higher system pricing than many developed PV markets and there is large opportunity for non-hardware cost reductions.
- Current tax benefits may be partially inflating U.S. system prices, therefore the impact of the expiration may not be as great as expected.
- There has been considerable progress in the past year introducing lower cost financing vehicles in the solar market. Securitized products, like the ones recently developed by SolarCity, Mosaic, and NRG, have already introduced sources of capital at a cost that is below many traditional sources of capital for the PV industry.
- The growth of PV market has also translated into more funds invested in the sector. The amount of funds annually invested in U.S. distributed solar projects has increased by around 85% since 2010, growing from \$4.3 billion to over \$8.0 billion in 2013, as shown in Figure 6.

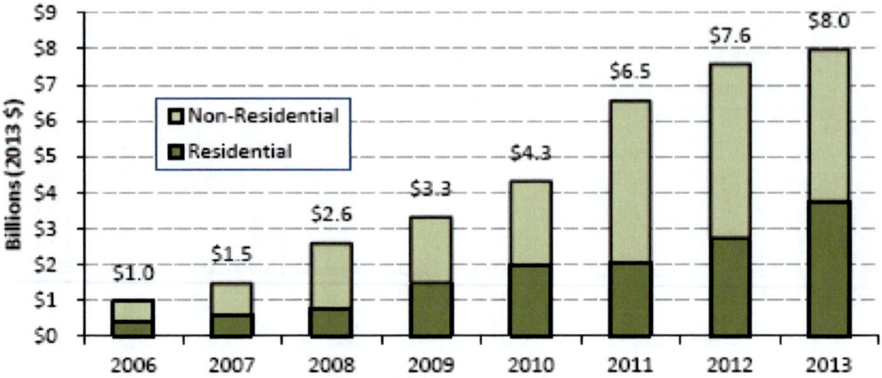


Sources: GTM & SEIA (2014); Barbose et al. (2011).

Figure 6. U.S. solar workforce & value of installed PV systems, by sector (annually).

If analysts' U.S. installation projections are correct, there will be $10 billion to $20 billion of available U.S. solar project investments in the next several years. Given the need for additional capital, there is great potential for lending institutions to fill a portion of this gap.

## 2.3. Solar Loans: A Nascent Option

Loan products designed to specifically finance solar installations have, historically, not been widely available in the U.S. market (Hawaii's range of solar loan offerings are a notable exception). This meant that home and business owners who chose ownership over third-party finance would have to either buy the system outright (i.e., a cash purchase) or finance through another, more general type of loan product such as a home equity line of credit (HELOC) or a standard commercial loan. Even before third-party ownership became more widely available in 2008, solar-specific loan programs were scarce.

However, several recent market changes have caused solar loans to gain market share. The U.S. solar market has become large enough to garner the attention of lending institutions. Additionally, with the housing market rebound and the overall improvement in the economy, banks have felt more comfortable lending to individuals and businesses. Finally, with the steady year-over-year declines in the costs associated with installing a solar system, the necessary size of the loans have decreased to the point that monthly payments are much more manageable. In 2017, with the reversion of the section 48 tax credit from 30% to 10% and the elimination of the 25D personal tax credit, many businesses and individuals may not need to rely on third-party providers to take advantage of the tax credit, and may be more likely to use a loan to finance solar instead.[4]

# 3. Solar Loans: Lenders, Processes, Products, and Risks

## 3.1. Lenders

Several types of financial entities can participate in the solar loan value chain. To categorize these entities, this report distinguishes between depository institutions—which comprise banks and credit unions—and non-depository institutions, namely finance companies and other types of capital providers (hedge funds, private equity, etc.).

### *3.1.1. Banks*

A bank is an institution that accepts deposits and channels these deposits into lending activities, making loans to its customer base. In this sense, a bank is a financial intermediary that allocates capital from savers (depositors) to those seeking capital (borrowers). The efficient flow of capital is a touchstone of most healthy modern economies, and banks serve this core function (Heffernan 2005; Gup 2011). Banks can offer products and services to both commercial businesses and individuals (retail).

As of October 2014, there were about 6,800 active banks in the United States, accounting for a total asset base of $14.7 trillion and deposits of about $9.4 trillion (ABA 2014). Smaller banks, commonly referred to as "community" banks, constitute over 95% of all U.S. banking institutions but manage less than 1/6 of total banking assets. The rest of U.S. banking assets are consolidated within 390 banks, with the top five managing over half of the $14.7 trillion (The Fed 2014a) in U.S. banking assets. As a comparison, U.S. gross domestic product— the total value of all U.S. goods and services—was about $16.8 trillion in 2013 (The Fed 2014b).

### *3.1.2. Credit Unions*

Credit unions, while outwardly similar to banks, are generally not classified as banking institutions. They are depository entities that offer loans and other financial products to their members, but they are distinguished by their not-for-profit tax status (which exempts them from federal income tax) and their organization around affiliations of federal, state, or corporate scope. There are, for example, credit unions for members of certain branches of the military, for members of certain trade unions, or for workers at certain companies (Heffernan 2005). Because of their tax advantaged status, credit unions are generally able to offer slightly higher deposit rates than those of traditional banks, as well as slightly lower interest rates on loans (NCUA 2014a).There are over 6,500 credit unions in the United States with assets totaling over $1 trillion. Credit union membership comprises over 97 million Americans (NCUA 2014b).

### *3.1.3. Non-Depository Private Entities Associated with Lending*

This category is comprised of a range of finance companies and capital providers that could or currently do participate in the solar loan value chain. These include not only solar companies such as SolarCity, Sungage Financial, Kilowatt Financial, OneRoof, and others, but also hedge funds, private equity, and other types of investors with large balance sheets.

Solar loan programs sponsored by finance companies cannot make use of depositor capital, and therefore require some other source to fund loan origination. These sources could be public or private and could be depository or non-depository. Common sources of capital include:

- Large banks, such as Bank of America or Citibank, which can provide aggregation facilities (also known as “warehouses”) to pool loans once they have been originated
- Hedge funds, private equity, or other types of managed pools of capital pools
- State-level entities, such as green banks or development authorities (see Connecticut and New York for examples of loan programs backed by state institutions)
- “Crowdsourced” capital or peer-to-peer loans facilitated through platforms such as Mosaic.

The cost of capital for depository institutions is typically much lower than that of the non-depository entities. Banks do not currently pay more than a few basis points on their checking and savings accounts, which means that they can access capital at nearly risk-free rates (Bankrate 2014). Non-depository institutions do not enjoy such an arrangement, and must pay rates on their capital that reflect the perceived level of risk associated with their business and balance sheet.

While banks may be able to draw on pools of lower cost capital to fund their loan portfolios, they do face regulatory costs to a degree often in excess of finance companies. For example: a bank's cost of capital (the rate it pays on deposits) averages around 1% and the interest rate it charges on solar loans is 6%. Instead of capturing all of the 5% difference between its cost of funds and the rate of return on its solar loans, the bank may be required to devote a portion of this difference to purchasing high quality assets to hold in reserve as per safety and soundness regulations (see Section 3.2).

Additionally, finance companies have options that can reduce the interest rate that they offer on their solar loans to levels below their own cost of capital. Such options include interest rate buydowns from the solar installer (referred to as "dealer discounts" in this paper; see Section 3.4), protracted loan maturities, and others.

### *3.1.4. Non-Depository Public Entities Associated with Lending*

Federal, state, and municipal governments and utilities may also participate in the solar loan process. These entities do not have access to low-cost capital through deposits (as do banks) but through taxpayer funds and through typically lower rates paid on debt relative to private corporations and individuals.

Public entities may allocate capital to solar loan programs through a variety of vehicles, including green banks, public benefit corporations (such as port authorities), credit enhancements, bonding, or through actual administration of a loan program. Property assessed clean energy (PACE) programs, for example, require participation from both state and local governments. The state legislature must first pass PACE-enabling legislation so that tax assessments can be placed on properties seeking to finance solar installations or energy retrofits. Adopting municipalities may either issue bonds to fund these improvement projects, or may administer the program if private lenders are participating. Additionally, the state may also participate in PACE lending through a green bank, as does Connecticut's Clean Energy Finance and Investment Authority (CEFIA), which administers the state's commercial PACE program and has originated loans. CEFIA also provides credit enhancements for certain financial transactions that fall under its mandate.

## 3.2. Regulation

Lending in the United States is a highly regulated business, though the regulations and regulators can vary considerably based on the geographic location and type of lender in question, as well as the markets in which that lender participates.

At one end of the spectrum is the banking industry, which is one of the most heavily regulated in the world. The stringency of the rules governing banks' capital requirements, asset mix, risk management, and other aspects of its business derives from a long history of bank failures throughout U.S. history. Large-scale bank failures can freeze the flow of capital in advanced economies, and the loss of consumer deposits can be deleterious to GDP.

For these reasons and others, banks are subject to heightened regulation relative to other industries. Three separate federal agencies are responsible for monitoring and ensuring the health of banks in the United States: the Office of the Comptroller of the Currency (OCC), The Federal Reserve Banking System (the Fed), and the Federal Deposit Insurance Corporation (FDIC). Each regulator oversees a different class of banks, though there is some overlap in scope between the three agencies (see Table 1). Additionally, banks can also be regulated at the state level, depending on their charter.

Banking regulations in the United States are largely focused on five areas of protection:

1. Ensuring safety and soundness to protect against potential bank failures and therefore losses to FDIC funds and potentially depositors
2. Insuring deposits
3. Ensuring capital adequacy, which, like safety and soundness requirements, ensures that banks have enough Tier 1 (i.e., very low risk) capital to weather economic shocks
4. Ensuring compliance with consumer protection laws
5. Mitigating the systemic risk of the largest financial institutions to the national and global economy (Jickling 2010).

Credit unions, while similar to banks in terms of products and services, do not fall under the supervision of the three U.S. bank regulators, and therefore are not necessarily subject to these same regulations. Credit unions have their own federal regulator, the National Credit Union Administration or NCUA, which does provide depository insurance and must examine and evaluate credit unions according to the riskiness of their assets. However, because credit unions are not considered to be systemically risky, they are not subject to the same stringent capital requirements and safety and soundness measures as banks (CRS 2010).

Non-depository finance companies fall under the purview of several regulators, depending on their location and activities in the marketplace. These regulators include: state lending authorities; the Securities and Exchange Commission (SEC) for entities that buy and sell securities; the Commodity Futures Trading Commission (CFTC), which is responsible for the oversight of derivative contracts and trading; and the Consumer Finance Protection Bureau (CFPB), which is the newly-created federal agency to guard against predatory financial practices that harm individuals.

**Table 1. U.S. Bank and Non-Bank Regulators**

| Bank Regulators | | | |
|---|---|---|---|
| Regulator | Regulated Entities | Authorities | Associated Laws |
| The U.S. Federal Reserve System (The Fed) | • Bank and other financial holding companies as well as some subsidiaries<br>• State banks that are members of the Federal Reserve System<br>• U.S. branches of foreign banks and foreign branches of U.S. banks | • Lender of last resort<br>• Control inflation and<br>• deflation<br>• Manage money supply; execute monetary policy<br>• Enforce capital requirements and assess risk-weighting | • Federal Reserve Act of 1913<br>• Dodd-Frank Act of 2010 |
| Office of the Comptroller of the Currency (OCC) | • National banks<br>• U.S. branches of foreign banks<br>• Federally chartered thrift institutions | • Enforce capital requirements andassess risk-weighting | • National Banking Act of 1863<br>• Dodd-Frank Act of 2010 |
| Federal Deposit Insurance Corporation (FDIC) | • Federally insured depository institutions<br>• State banks that are not members of the Federal Reserve System<br>• State-chartered thrifts | • Depository insurance<br>• Capital requirements<br>• and risk-weighting | • Glass-Steagall Act of 1933<br>• Banking Act of 1935<br>• Dodd-Frank Act of 2010 |

**Table 1. (Continued)**

| Non-Bank Regulators | | | |
|---|---|---|---|
| Regulator | Regulated Entities | Authorities | Associated Laws |
| U.S. Securities and Exchange Commission (SEC) | • Securities exchanges<br>• Brokers and dealers<br>• Clearing agencies<br>• Mutual and hedge funds<br>• Ratings agencies<br>• Financial entities that buy and sell securities | • Oversight of U.S.<br>• securities markets<br>• Enforcement of U.S.<br>• securities law<br>• Set financial accounting<br>• standards | • Securities Exchange Act of 1934<br>• Securities Act of 1933<br>• Trust Indenture Act of 1939<br>• Investment Company Act of 1940<br>• Investment Advisers Act of 1940<br>• Sarbanes-Oxley Act of<br>• 2002<br>• Dodd-Frank Act of 2010 |
| Commodity Futures Trading Commission (CFTC) | • Futures exchanges<br>• Financial entities that buy and sell swaps and other derivative contracts<br>• Derivative execution facilities and clearing organizations | • Police derivative markets for abuses | • Commodity Futures<br>• Trading Commission Act of 1974<br>• Commodity Exchange Act of 1936<br>• Dodd-Frank Act of 2010 |
| Consumer Finance Protection Bureau (CFPB) | • Banks, credit unions, credit card companies, loan servicers, and other financial companies in<br>• U.S. consumer finance | • Write rules, supervise companies, and enforce federal consumer financial laws | • Dodd-Frank Act of 2010 |

Source: Jickling et al. (2010).

## 3.3. Processes and Business Models

The lending value chain, regardless of asset class, can include the following general steps/roles:

- Customer Acquisition: The process of generating leads and closing the sales process.

- Loan Approval and Origination: The process of evaluating potential customers' credit profile and determining whether it meets the standards set by the lender (according to their particular underwriting criteria). If a particular customer's credit is approved, then the loan will be originated, i.e., disbursed to the appropriate party according to the terms of the loan.
- Servicing: The servicer maintains the collection accounts associated with a particular loan portfolio and is responsible for ensuring the borrowers' repayment.
- Aggregation/Warehousing: Loans are booked as assets on the balance sheet. The loan originator has the option to aggregate and hold these assets, or sell them down to a warehouse facility if they have secured one with a large bank or financial institution. A warehouse facility is essentially a fund which "buys" the loan assets from the originator (typically at a discount), allowing the initial loan holder to free up balance sheet capacity and originate additional loans.
- Securitization: The loan holder—i.e., the originator, the warehouse provider, or some other buyer—may securitize the loan portfolio, provided that it is of sufficient volume, and the anticipated credit rating on the transaction will allow the issuer to earn a favorable rate on the resulting debt. It is worth noting that the financial assets derived from loan securitizations—known as collateralized loan obligations (CLOs)—have been in lower demand since the financial crisis. Accordingly, while securitization is still an option for some loan holders, it may not be a primary consideration (Lowder et al. 2013).

The value chain specific to solar loans includes the following asset-level steps/roles:

- *Installation:* This encompasses all the physical processes of the PV system construction on the borrower's rooftop.
- *Operations and Maintenance:* While all TPO contracts include operations and maintenance (O&M) of the solar system for the life of the contract, there is currently no such standard offering in the loan product marketplace. Some loan programs offer O&M services bundled into the loan amount (e.g., Mosaic and Dividend Solar), though with others (Admirals Bank, most PACE programs), O&M remains the responsibility of the system owner. O&M services in the

loan market reportedly do not feature the same extent of coverage as do those in TPO financings.[5]

Much like in the TPO market, solar loan business models can either tend toward aggregation (i.e., a single company fulfills several of the functions along the loan value chain) or disaggregation (functions along the value chain are mostly fulfilled by strategic partners). A vertically integrated company like SolarCity provides an example of the aggregated model, where the customer acquisition, underwriting/origination, installation, O&M, loan aggregation, and securities issuance could all be performed in-house or through subsidiaries.

Conversely, Sungage Financial provides an example of the disaggregated model. Sungage performs customer acquisition, loan origination, and servicing in-house, but relies on strategic partners to fulfill other aspects of the value chain. Installation is performed by vendors in Sungage's dealer network, and these installers also provide workmanship warranties as a sort of O&M service. Sungage Financial operates an origination platform through which it aggregates demand and develops portfolios of loans that are capitalized by third-party investors.

## 3.4. The Solar Loan Product Landscape

Table 2 below offers a general outline of the various forms that a solar loan may take in the marketplace with examples of currently available programs.

**Table 2. Overview of Solar Loan Options Available in the 2014 Market**

| | *Loan Type* | *Description* | *Lender* | *Term (years)* | *Interest Rate* | *Examples* | *Availability* | *Security Interest* |
|---|---|---|---|---|---|---|---|---|
| Residential | Solar-Specific | A loan made to a borrower exclusively for the purchase of a solar system | Bank, CU*, Finco** | 5 – 20 | 2.99% – 8% + | •EnerBank •GreenSky •Mosaic •Sungage Financial | Can be limited to certain jurisdictions or available nationwide | Solar assets (i.e., the system) or unsecured |
| | Title I (HUD) | 90% government-guaranteed loan secured by second | Program-approved Lenders (Bank, CU, | 10 – 20 | 3% – 9% | •PowerSaver •Admirals Bank | Nationwide | Lien on thehome (second mortgage) |

| | *Loan Type* | *Description* | *Lender* | *Term (years)* | *Interest Rate* | *Examples* | *Availability* | *Security Interest* |
|---|---|---|---|---|---|---|---|---|
| | | lien on home | Finco) | | | | | |
| | On-Bill | Loan amortized through electric bill | Bank, CU, Utility, State | ≤15 | 3% + | NYSERDA | Service territories of participating utilities | Utility bill |
| | Home Equity | A loan or line of credit against the value of a home-owner's equity. Also called a 2nd mortgage | Bank, CU, Finco | ≤30 | 4.5% + | N/A | Nationwide | Lien on thehome (second mortgage) |
| Both | PACE | A loan made via an increase in property tax assessment and amortized through property tax bill | Bank, Finco, Muni-cipality | ≤20 | 5% – 8% | •Hero Program •Renewable Funding | States with PACE-enabling legislation | Tax lien |
| Commercial | Comm. Loan | A loan made to businesses to finance operations, capital expenditures, etc. | Bank, CU, Finco | 5 - 30 | 3.25% – 6.85% | NA | Nationwide | Borrower's balance sheet |

* CU: Credit union.

** FINCO: financial company.

It is worth noting that some solar loan programs on the market today make use of a "dealer discount" to reduce the interest rate on a borrower's loan. The dealer discount is a sum of money paid by the solar installer to the lender on a per loan basis, and which functions as an interest rate buy down. In this way, some products are available at APRs of as little as 2.99% with amortization schedules from 10-15 years.

While the dealer discount does look attractive from a borrower standpoint, it could present some regulatory complications. Legal guidance differs here, but NREL's conversations with lenders indicate that they must, per consumer protection laws, disclose the difference in price between a system financed with a dealer discount (which would be higher to account for the dealer's contribution), and one where no such discount was applied (which would be closer to the benchmarked per-watt cost in that market). There is still a degree of uncertainty as to the best practices in dealer discounts, though with the volume of loan products coming on the market making use of this feature, some clarity will likely be forthcoming.

## 3.5. Challenges in Lending to the Solar Asset Class

Investor understanding of the solar asset class is improving (recent securitizations and high deal volumes over the last several years have contributed to this outcome), though loan products in particular may have to contend with a number of outstanding risks. These include: a lack of historical performance and credit data (this is an issue for other types of solar financings as well); regulatory treatment of solar loans; complications with the characterization of solar assets as fixtures versus personal property (depending on whether they are financed via a lease/PPA or a loan); the rights of secured-solar lenders in foreclosure situations; and other issues without experiential learning or legal precedent may present some barriers to the growth of the solar loan market.

Banks in particular face a number of regulatory and legal challenges associated with the solar asset class. At current system prices, PV assets typically necessitate a longer amortization schedule to ensure that monthly payments are not too burdensome for customers. However, long term loans—greater than five years—can carry considerable interest rate risk, whereby a rise in market rates can negatively impact the bank's income on a fixed-rate asset (such as a solar loan). This can either occur through the opportunity cost of holding a loan that generates interest income at below the new market rate, or by shrinking the spread between a bank's cost of capital and its return on capital until the loan reaches maturity. There is an option for banks to hedge against interest rate risks by the use of derivatives such as interest rate swaps or caps, but these strategies may be beyond the level of sophistication at which many banks are comfortable. Hedging, moreover, does come at a cost.

Additionally, regulators may also review a bank's loan portfolio unfavorably if its portfolio is heavily weighted in un-hedged, long-dated assets, or if the portfolio is over-hedged. Hedges and loans with long maturities have higher risk weightings, and federally insured banks—per regulatory requirements codified in the 2010 Dodd-Frank legislation—must hold certain ratios of Tier 1 (i.e., high quality or risk free) capital against risky assets. This Tier 1 capital, while contributing to the safety and soundness of the financial institution, cannot be "put to work" in the market, thus limiting the ability of banks and other investors to direct resources to emerging markets such as solar. And because banks tend to be relationship-driven, they may allocate their increasingly scant capital to existing customers instead of opportunities in perceived riskier markets.

# 4. Economics of Financing a PV System Through a PPA versus a Loan

This section compares the economic attractiveness, to an individual or business, of using a loan versus a PPA to finance a PV installation. It also discusses non-economic factors that might influence an individual or business in their decision making. The results of the economic models do not demonstrate which option is the most efficient for every circumstance; rather, the analysis and subsequent discussion attempts to highlight some of the key factors a company or individual should think about in determining which financing method is best for their specific circumstance.

## 4.1. Pro Forma Financial Models

### *4.1.1. Pro Forma Financial Model Description*

To compare the economic attractiveness of financing residential and commercial PV systems through a typical loan product with a third-party power purchase agreement (PPA) the authors built three pro forma financial models: 1) a residential/commercial PPA model; 2) a residential loan model with varying tenors; and 3) a commercial loan model with varying tenors. Each model solves for the LCOE that satisfies all assumptions outlined in Table 3. For a more detailed set of assumptions, see Appendix A.

**Table 3. Basic Model Assumptions**

| Assumption | Residential | Commercial |
|---|---|---|
| Installed Price | $3.74 per watt | $2.39 per watt |
| Location | San Diego, CA | San Diego, CA |
| Size | 5 kW | 500 kW |

**Table 3. (Continued)**

| Assumption | Residential | Commercial |
|---|---|---|
| Project Lifetime (PPA length) | 20 years | 20 years |
| Incentives | 30% Federal Tax Credit | 30% Federal Tax Credit, 5-year MACRS depreciation schedule |
| Loan Interest Rates | 5 year: 6.00% 10 year: 7.00% 20 Year: 8.00% | 5 year: 3.75% 10 year: 5.20% |
| 3rd-Party Capital Cost (after-tax) | 9.2% return (9.0% tax-equity; 10.5% sponsor equity) | 9.2% return (9.0% tax-equity; 10.5% sponsor equity) |
| Host Business or Individual's Capital Cost | 6.2% after-tax return (8.7% pre-tax) | 10.0% pre-tax return (6.0% after-tax) |

In effect, LCOE was used to provide comparisons between models, acting as a proxy for the local utility retail rates necessary to satisfy the return requirements of both investors and customers. Annual payments during the term of the contract were also calculated in order to estimate how each scenario affects a home or business owner's cash flow.

All models assume system location in California. Not only is California the largest U.S. solar market, but it is also economically feasible to install systems there without state incentives, allowing for the exclusion of this complicating factor from the models. Systems were assumed to have a 20-year *economic life* to align with the typical term length of a PPA. The pro forma models also calculate cash flows in years 20 through 30 to measure additional benefits beyond the life of the contract, but within what is typically considered the lifetime of a PV asset.

The discount rates for residential and commercial customers were assumed to be 6.2% after-tax (8.7% pre-tax) and 6.0% after-tax (10.0% pre-tax) respectively, based on assumptions and calculations outlined in Appendix A. Although third-party providers might have additional costs (mostly attributable to the cost of capital associated with deploying TPO systems) (Feldman 2013b), we assume that system price and operating expenses are not affected by differences in financing method.

#### *4.1.2. Residential/Commercial PPA Pro Forma Financial Model*

In the PPA pro forma model, the contract price per kWh given to residential and commercial customers was determined by the calculated

LCOE, reduced by the customer's discount rate (i.e., PPA = LCOE x [1 − discount rate]. The commercial discount rate was increased by the commercial entity's tax rate so as to provide it with an after-tax return, comparable with self-ownership using a loan (i.e., $\text{PPA} = \text{LCOE} \times \left[1 - \frac{\text{discount rate}}{1 - \text{tax rate}}\right]$).

It was assumed that projects were financed using a sale-leaseback transaction between the developer (sponsor) and a tax-equity provider.[6] The sponsor funds 15% of the project cost (in the form of an upfront lease prepayment, which the sponsor expenses over the contract lifetime). The sponsor is required to make yearly lease payments to the tax-equity provider and to pay operating expenses associated with the project. In exchange, it receives all PPA revenues from the system host. Based on industry data, the sponsor is assumed to require a 10.5% after-tax rate of return. For a more detailed set of assumptions, see SolarCity's calculated cost of capital in Appendix A. The tax-equity provider funds 85% of the asset, net of lease prepayments, and in exchange receives lease payments from the developer, and an investment tax credit and 5-year modified accelerated cost recovery system (MACRS) depreciation associated with the project. Based on industry data, the tax-equity provider is assumed to require a 9.0% after-tax rate of return (Martin 2014). In the model, the developer leases the system for 80% of the term of the contract, at the end of which it purchases the system back from the tax-equity investor for 20% of the original installed price. This ensures that the arrangement qualifies as an 'operating' rather than 'capital' lease.[7] The net cash flow providing the return to each participant is calculated on an after-tax basis.

#### *4.1.3. Residential Loan Pro Forma Financial Model*

In the pro forma model, the residential homeowner obtains a loan for the full cost of the project, but immediately pays down 30% of the principal in the first year with the benefits of the 30% investment tax credit. While the homeowner receives no revenue from the PV system, its yearly benefit is calculated as the electricity expense savings, or the calculated LCOE (i.e., utility retail rate) multiplied by the expected PV system production. LCOE is adjusted in the model to provide the homeowner with an 8% return on cash flows (net of loan payments, O&M, and energy savings). The homeowner is responsible for all operating expenses, including principal and interest payments on the loan. While the interest on certain loans currently available in the marketplace is tax deductible, it is not assumed in this model; the assumption would further reduce the calculated LCOE. As the PV system is built for personal use, the homeowner does not deduct any depreciation expense, as in the case of a commercial enterprise. However, also unlike a commercial entity, the reduced electricity expenses occur on an after-tax basis (i.e., the benefit is not reduced by the tax rate). The residential loan model creates pro formas for 5-, 10-, and 20- year debt terms to reflect the range of loan products currently available in the marketplace. Longer tenors are generally harder to finance, require more paperwork, and have higher interest rates due to the increased risk. Based on market data and private conversations with industry, the interest rates used for this model were 6%, 7%, and 8% for the 5-, 10- and 20-year debt terms, respectively (Admirals 2014).

#### *4.1.4. Commercial Loan Pro Forma Financial Model*

In the commercial loan pro forma model, the business obtains a loan for the full cost of the project, but pays down 30% of the principal within the first year because they also receive a 30% investment tax credit in that time frame. Also like the homeowner, the business receives no revenue from the PV system, so their yearly benefit is calculated as electricity expense savings (the calculated LCOE multiplied by the expected PV system production). The LCOE is adjusted to provide the business with a 9% return. The business is responsible for all operating expenses, including principal and interest payments on the loan. It also expenses the cost basis of the PV system using the 5-year MACRS schedule, as well as the interest associated with the loan. However, unlike a homeowner, because utility expenses reduce the operating income of a commercial entity, the utility bill savings from PV produce higher operating income and thus higher taxes. The model captures this impact as well.

Commercial entities typically borrow money under shorter time horizons than residential homeowners, including the loan products available for PV installations. Therefore, the pro formas created in the commercial loan model use 5- and 10-year debt terms. Based on private market data, the interest rates used were 3.75% and 5.20%, for the 5- and 10- year debt terms, respectively.

### 4.2. Results of Residential Pro Forma Financial Models

The results of the pro forma financial models show that using either the PPA or bank loan to finance a residential PV system offered savings over utility retail rates (based on San Diego Gas and Electric's [SDG&E] blended residential rate for Tiers 2 and 3[8]) as demonstrated in Figure 7. However, the cost of energy for the bank loans was lower than the cost of energy for a PPA by 19% to 29% (varying by the term of the loan), due to the higher cost of capital necessary for the sponsor and tax equity in a PPA transaction. Similarly, because the interest rates for the shorter loans were lower than longer term loans they were also able to provide a lower cost of energy.

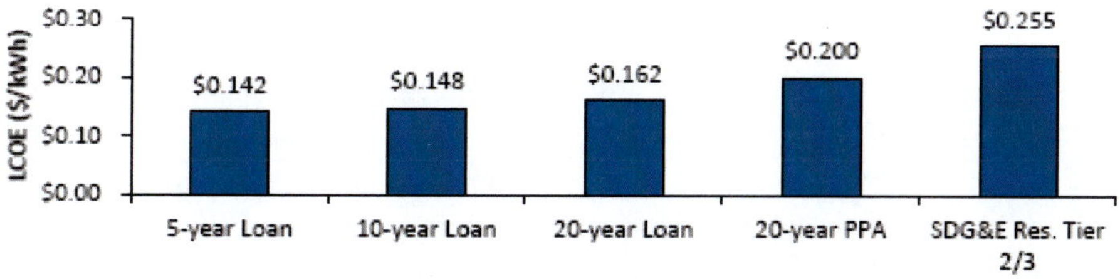


Figure 7. LCOE of residential PV systems, financed under a PPA or loan, compared to retail rates.

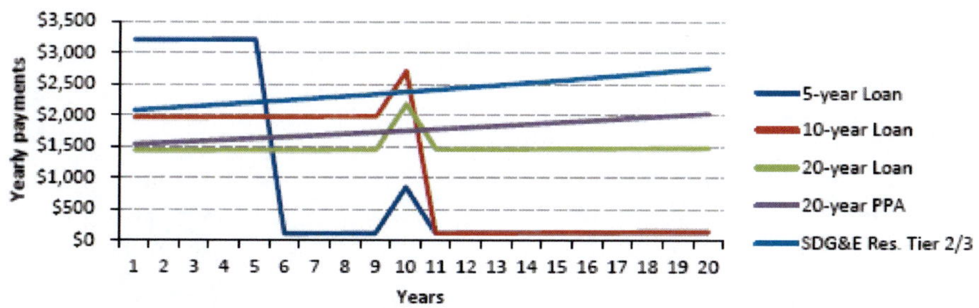


Figure 8. Comparison of annual payment between residential PV systems, financed under a PPA or loan, and retail rates.

However, while the 5- and 10-year loans resulted in a lower cost of energy over the life of the contracts, annual principal and interest payments to service these loans were typically higher than the annual payments for the longer loans, the PPA, and the price a customer would have paid the utility for the same amount of energy – as demonstrated in Figure 8. In fact, in the first year of a 5-year loan, the debt service payments were calculated to be more than 50% of what a customer would pay to a utility, and over twice as much as a PPA or 20-year loan. At the end of the loan term, a customer's annual payments would drop to the cost of O&M, including an inverter replacement in year 10, however, some customers may not want to wait five or ten years to start saving money. Additionally, a customer may not want the liability of a potentially high cost of repair or replacement to the system, such as the assumed inverter replacement in year 10. A PPA customer has a hedge against the cost of electricity, including the cost of maintaining the solar system.

## 4.3. Results of Commercial Pro Forma Financial Models

Commercial retail electricity rates vary dramatically depending on customer load profile. For this reason, we did not develop a commercial retail rate comparable to a PPA or loan product. However, as demonstrated in Figure 9, LCOE for the bank loans was 43% to 47% lower than the cost of energy for a PPA (varying by the term of the loan) owing to the higher cost of capital necessary for the sponsor and tax-equity provider in a PPA transaction. Additionally, unlike in the residential pro formas, the commercial models calculated the cost of energy under the 10- year loan to be 5% lower than the 5-year loan – despite the fact that the longer loan has a higher interest rate. This result is caused by the interest deduction from the loans, and the tax-rate and discount rate of the commercial entity; these factors make it more beneficial to pay more interest, cumulatively, but have less cash outlay in earlier years, net of taxes.

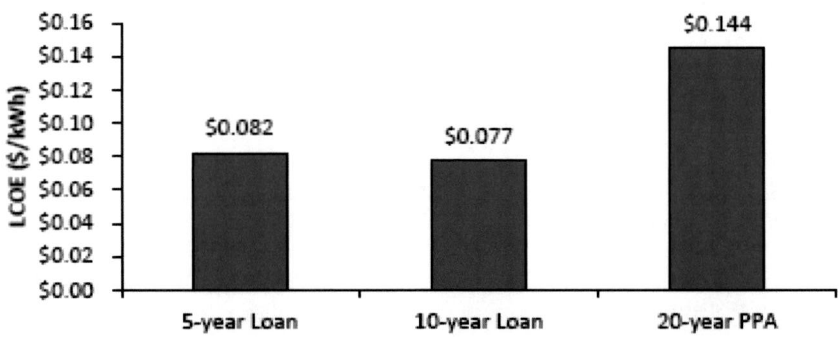


Figure 9. LCOE of commercial PV systems, financed under a PPA or loan, compared to retail rates.

As shown in Figure 10, a 5-year loan would result in higher annual payments (for the first five years, almost double) compared to the alternative commercial pro forma models. However, while the PPA and retail rates increase with inflation (or PPA escalator), the loan has a fixed payment over time (other than O&M) – so that in the first ten years the annual payments made under the loan are sometimes less than it would have paid to the PPA provider (and much less afterwards).

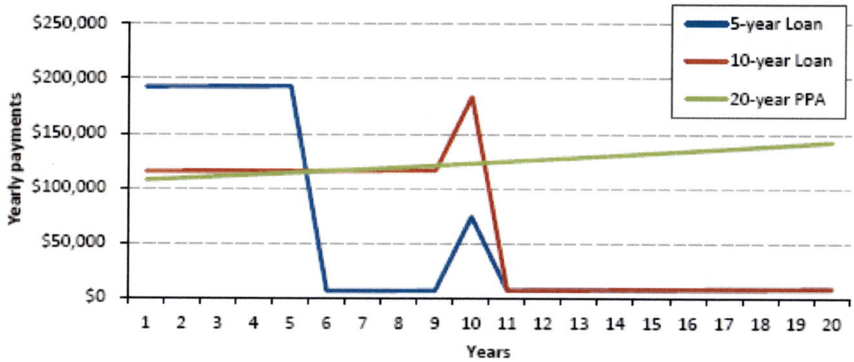


Figure 10. Comparison of annual payment between commercial PV systems, financed under a PPA or loan, and retail rates.

## 4.4. Economic Factors Beyond the Pro Forma Financial Models

The pro forma models demonstrate that debt products' lower cost of capital provides more economic benefit to commercial and residential customers than the modeled higher-cost PPA, which relies on sponsor equity and tax equity. This result, however, depends on the assumed rates of return for financiers, which might vary for several reasons. For example, in Admirals Bank Federal Housing Administration (FHA) Title I Home Improvement Loan, interest rates are dependent not only on term (as discussed previously), but on a customer's qualifying FICO score (i.e., credit risk). In recently published documentation (as of this writing), interest rates on Admirals Bank 20-year loan varies from 5.95% (approximately 200 basis points below the pro forma assumptions) for customers with FICO scores between 795-850, to 9.95% (approximately 200 basis points above the pro forma assumptions) for customers with FICO scores between 650- 659 (Admirals 2014). Adjusting the interest rate by 2% changes the calculated pro forma model's LCOE by $0.02-$0.03/kWh in either direction.

Additionally, this doesn't account for the 10% to 20% of the population with a credit score between 580 and 650 (Huynh 2014), meeting the FHA minimum FICO credit score to receive a home loan (though, credit quality may also prove problematic to participate in a PPA program as well). These lower FICO scores could result in loans with interest rates upward of 10% unless the potential customers receive one through a state-sponsored program.[9]

PPA providers are also trying to lower their cost of capital through innovative financing mechanisms, such as securitizing project portfolios. SolarCity, a residential and commercial PPA provider, recently raised funds from three securitizations, at coupon rates between 4.0% and 4.8%, backed by cash flows from pools of its distributed PV systems – capital that could go to fund future PV systems. Although the 4.0%-4.8% rates are not exactly comparable to the required returns in the model, they do represent a significant reduction in the cost of capital normally required for TPO financing of a portfolio of PV assets.

Additional factors might make a loan more attractive than a PPA. Homeowners are sometimes able to claim the interest payments on several loan products—including some used to finance a residential PV systems—as a tax deduction (e.g., home equity loan). These deductions would provide an even larger economic return, thus lowering the cost of energy.[10]

Our modeled PV systems have an economic life of 20 years, while the average PV system is expected to last for at least 30 years.[11] A PV host that finishes the term of a loan owns the system outright. At the end of a PPA contract, however, the host must purchase the system for fair market value, sign a new PPA, or allow the system to be removed from its property. We estimated the additional value of ownership in years 20-30, by comparing the LCOE of the 20- year loan pro formas to that of the 30-year loan pro formas. See Figure 11.

Figure 11. LCOE comparison of PV systems, financed by loans, over 20 and 30 year lifetime.

Based on the calculations represented in Figure 11, extending the life of the pro forma to account for the full life of a PV asset lowers the LCOE by an average of 18%.

Although our models assume that system costs remain the same regardless of financing structure, loans typically involve fewer parties in the transaction than do PPAs, thus might be able to build and finance a system at a lower price. In addition, TPO systems often require the use of industrial-grade electric meters, which may add costs relative to the meters needed for a host-owned system. Feldman et al. (2013b) estimate that, "Third-party-ownership-related costs add $0.78/W to a residential portfolio and $0.67/W to a commercial portfolio."

However, there are also many reasons companies or individuals may find a PPA contract more beneficial than using a loan to purchase the system themselves. As noted earlier, the term of a loan is one factor in determining the size of annual principal and interest payments. While a system owner may be able to receive a lower interest rate from a shorter-term loan, the higher annual payments during the tenor of the loan may prove cost prohibitive (or unattractive). This is particularly true for agents with time horizons shorter than the system. For example, a company's facilities manager, who is unlikely to remain in the same job for the economic life of the PV system, might choose immediate energy savings over potential long-term economic value. A company might also require immediate savings to approve a project. Additionally, even if the loan is the same term as a PPA (which is often not the case because long-term loans are currently hard to find), the fixed nature of a loan payment over the term of the loan may cause the loan payments in earlier years to be larger than the utility savings received relative to utility rates (or PPA payments).

Commercial entities must also consider how various financing options impact their financial statements. While PPAs and system purchases are both long-term commitments, PPAs are considered off-balance sheet transactions; similar to an operating lease, PPA payments are treated as operating expenses, and the long-term liability of the contract does not appear on a company's balance sheet. Financial statements are used to measure the financial health of a company, by internal and external parties, and corporate debt often has covenants that limit the amount of additional debt a company can incur. Therefore, a company might be unable to add the liability of a PV loan onto its balance sheet and might instead opt for a PPA. Much of this depends on the size of the company relative to the amount of PV it aims to deploy.

Another consideration that affects both companies and individuals is the need for tax liability to take advantage of a significant benefit of system ownership. Currently both individuals and companies receive 30% of the cost of the system back in the form of a tax credit. Additionally, a company can also depreciate the cost basis of a PV asset using a 5-year MACRS depreciation schedule – potentially worth up to an additional 25% to 30%, depending on the corporation's tax rate. While these represent significant benefits, an owner must have sufficient taxable income to take advantage of them. Though the benefits can be carried forward to be used in later years, their worth in later years diminishes due to the time-value of money.[12] Based on average system pricing, as reported in Figure 4, a 30% ITC for a 5 kW system built in 2013 would require approximately $5,500 in tax liability; down from $9,500 for a 5 kW system built in 2010. According to data from the Congressional Budget Office (CBO 2013), as show in Figure 12, over 60% of U.S. households pay more than $5,300 in annual federal taxes, while 40% pay more than $9,700.[13] As the price of systems continues to fall, the number of households that can use the ITC in year one should increase. Businesses also pay varying degrees of federal taxes, though the degree to which an individual business can use a credit largely dependents on business size, PV system size (e.g., 5 kW, 5 MW), and the business' ability to adjust incurred income.

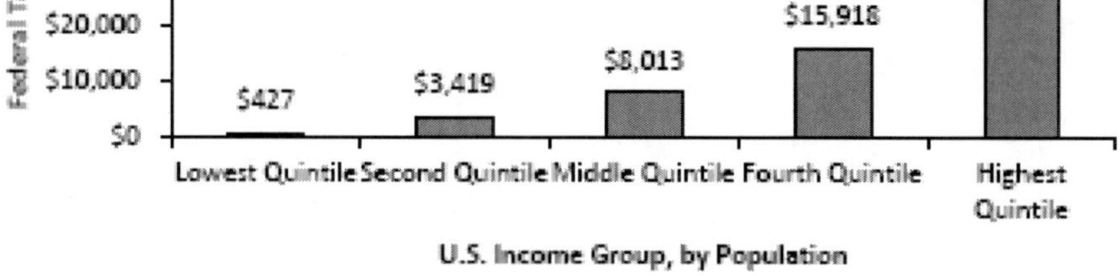


Source: CBO (2013).

Figure 12. Average U.S. federal taxes incurred in 2010, by quintile.

Depreciation is also an important difference when considering residential PV system ownership versus a third-party PPA provider. While both are able to claim a 30% investment tax credit (under Section 48 for a business and Section 25D for a homeowner), a business is also able to expense the cost basis of PV asset, using the 5-year MACRS schedule, which can offer a benefit worth an additional 30% of the cost of a system. While the ability to depreciate an asset, particularly at such an accelerated schedule, is a significant advantage, the corresponding disadvantage a PPA provider has compared to a homeowner is that they must pay taxes on any income generated from a PPA; in contrast, the electricity savings a homeowner experiences from a PV asset are all after-tax (i.e., an individual cannot deduct the cost of her personal utility bill on her taxes). The net gain or loss between the benefits of depreciation and the additional taxes depends on several factors, including the discount-rate and tax rate of the potential third-party system owner, the discount rate of the homeowner, the price of the PPA, the avoided cost of electricity, and the price of the PV system.

Businesses or individuals that own PV and those that finance it through a PPA also face different risk and reward considerations.[14] To a business or individual with a PPA the only risk of unexpectedly lower PV generation is the additional cost of electricity that must be sourced from the utility. In contrast, a PV owner must pay for the system, including O&M, regardless of system performance unless there is a production guarantee. If state and/or local incentives are required to make a system economical, system owners may also be exposed to issues such as renewable energy credit (REC) price volatility and delays in disbursements from grant programs. Further, there is a risk of interconnection delays and lost production while waiting for the utility to approve the system's operation.[15]

A system owner might also incur unforeseen O&M costs. For example, inverters do not typically have warranties for the full life of a PV system (which is why our model assumes an inverter replacement in year 10). While a system with a PPA may similarly require an inverter replacement in year 10, this cost is built into the PPA price. A system owner, on the other hand, must pay all O&M costs and so, in any given year, might incur costs higher than would have been incurred using a PPA.

Some loan programs currently on the market feature quality assurance components, which could mitigate some of the more significant O&M risks. For example, the solar finance company Mosaic and the inverter manufacturer Enphase have recently partnered to create a loan program in which Mosaic serves as the originator and initial loan holder, and Enphase provides O&M services over the 20-year life of the loan. The cost of this service is built into the principal amount, which gets amortized over the loan term. Another solar loan originator, Sungage, currently requires its installers to make production guarantees, so that system owners can have a reasonable expectation of system functionality through the loan term. While this is not a full O&M service, it does ensure some form of quality assurance, which could reduce the incident of defaults across Sungage's loan pool (assuming installers remain in business to honor the guarantees).

Many of these risks of PV ownership are small (such as system under production) can be hedged against (for example, by selling long-term REC contracts), or can be borne by another party (for example, by requiring a production guarantee from the installer or delaying payment for the system until receipt of state rebates and interconnection). Regardless, customers likely must educate themselves about PV risks and benefits more when they are buying a system than when they are entering into a PPA or other TPO contract.

## CONCLUSION

In June of 2014, GTM Research released its second update on the residential solar financing landscape. The report predicted that the TPO market, which has been a large driver of residential PV deployment for the last several years, would peak in 2014 and gradually cede share to loans and alternative forms of financing (such as PACE) thereafter (Litvak 2014).

While the actual competitiveness of the loan option in solar finance will be determined by the offerings in the market, this report attempts to provide a framework for understanding how loan structures could affect the ultimate cost to distributed PV consumers. Solar loans have the potential to provide an economical option (from an LCOE perspective) for homeowners and businesses to finance the purchase of a solar system, retaining the benefits of ownership that TPO systems cannot provide while avoiding the large upfront cost of a PV system. According to the analysis, solar loans at 5-, 10-, and 20-year maturities delivered a lower LCOE than a 20-year TPO PPA, representing a reduction of 19% to 47%. Both loans and TPO were found to generate electricity at a rate lower than the blended rate for SDG&E's second and third tiers of residential rates.

It is important to emphasize that while solar loans may result in a lower cost of generation to the consumer, thus reducing the lifetime costs associated with the system, the monthly payments on a loan may not necessarily reflect these favorable economics. Depending on the interest rate, the principal and the tenor of the loan, monthly P&I could be higher than the system owner would pay under both a TPO arrangement and the utility rate. However, the faster the loan is paid down, the more "free" electricity the host will enjoy post-financing, offsetting utility bills at no additional cost for the remaining useful life of the system. This becomes something of an optimization exercise for consumers deciding on a financing option, who may weigh their ability to make possibly higher monthly payments against the ultimate savings a loan could provide over the system lifetime. From a qualitative survey of the solar-specific loan options available in the market or in the planning phase as of this writing, NREL has determined that most products fall into the range of 5 to 20 year maturities. Of note, SolarCity has recently announced that it intends to roll out a 30-year product (Wesoff 2014), which would match typical mortgage maturities and could reduce monthly payments to a level highly competitive with or lower than utility rates in markets with lower power prices.

There are other factors, beyond loan tenor, that may impact the economics of a solar loan and TPO: the hosts' requirement for a tax liability to take advantage of the federal tax incentives; the additional liability and maintenance costs associated with ownership; the complications of adding assets and liabilities onto one's balance sheet (as opposed to a TPO transaction which is an off-balance sheet); and the economic time horizon of individual decision makers. Individual decision makers will have to determine how these risks and benefits compare with their own economic profile.

A broader range of loan financing options with flexible interest rates and maturities, coupled with deeper market penetration, could help reduce the financing costs associated with installing solar, and thus provide another trajectory for the achievement of the U.S. Department of Energy's SunShot goals. Distributed solar continues to perform robustly in terms of both growth and its ability to attract investment; the wider availability of financing, the diversity of financing options, and the competitive rates at which to finance in this space, will all prove influential to the ultimate success of this market, both in the near term and post-ITC.

# APPENDIX

### Table A-1. Basic Assumptions for Residential and Commercial Pro Forma Financial Models

| Installed Price ($/W) | Residential $3.74a | Commercial $2.39a |
|---|---|---|
| Capacity Factor | 18.6%b | 17.3%c |
| Inverter replacement ($/W) | $0.15d | $0.13d |
| Combined Federal & State Tax Rate | N/A | 40.2% |
| Annual Degradation Rate | 0.50%e | 0.50%e |
| Project Lifetime (PPA length) | 20 years | 20 years |
| O&M (% of installed cost) | 0.50%f | 0.50%f |
| Inflation | 2% | 2% |
| Federal Tax Incentives | 30% Investment Tax Credit | 30% Investment Tax Credit & MACRS Depreciation Schedule |

[a] Source: SEIA 2014b.

[b] Capacity factor calculated by PVWatts (version 1) for a PV system in San Diego, CA, with a derate of .84, azimuth of 180 degrees, and tilt of 25 degrees.

[c] Capacity factor calculated by PVWatts (version 1) for a PV system in San Diego, with a derate of .85, azimuth of 180 degrees and tilt of 5 degrees.

[d] Note: assumes that when inverter replacement occurs in year 10 of the project, prices have dropped targets outlined in The SunShot Vision Study, adjusted for inflation (see DOE 2012). Also assumes that the new inverters have a lifespan of 20 years, as the technology will have improved in the future.

[e] Source: Bolinger 2014.

[f] Source: Shah et al. 2013.

### Table A-2. Assumptions Specific to Loan Pro Forma Financial Models

| | Residential | Commercial |
|---|---|---|
| Percent Loan Coverage of Project Cost | 70% - remainder through tax credit | 70% - remainder through taxcredit |
| 5-Year Loan Interest Rate | 6.00% | 3.75% |
| 10-Year Loan Interest Rate | 7.00% | 5.20% |
| 20-Year Loan Interest Rate | 8.00% | N/A |

Source: based on market data and private conversations with industry.

## Table A-3. Assumptions Specific to PPA Pro Forma Financial Models Using an All-Equity Sale-Leaseback Transaction for Residential & Commercial Projects

| | *Tax Equity* | *Sponsor Equity* |
|---|---|---|
| *Ownership Percentage* | 85%[a] | 15% |
| *After-tax Rate of Return* | 9.0%[b] | 10.5%[c] |
| *Lease Length* | 16 years[d] | |
| *Residual Value* | 20%[e] | |

[a] Source: Bolinger 2014.

[b] Source: Martin 2014.

[c] After-tax cost of capital for SolarCity as calculated by the Capital Asset Pricing Model (CAPM); see below for calculations.

[d] Source: Peterson 2012.

## Table A-4. Assumptions for SolarCity's Cost of Capital Using Capital Asset Pricing Model (CAPM)

| Variable | Value | Explanation |
|---|---|---|
| Risk-free Rate (1) | 3.6% | 20-Year Treasury Bill Constant Maturity. December 1, 2013. Source: FRED 2014. |
| Equity Beta (2) | 1.88 | SolarCity stock beta. Adjusted close price of SolarCity stock for the time period 12/13/12 – 3/24/14. Source: finance.yahoo.com |
| Debt/Asset Ratio (3) | 46% | Source: SolarCity 2013 10K. |
| Tax Rate (4) | 0% | As of 2013, SolarCity had not yet earned taxable income. |
| Equity Risk Premium (5) | 5.3% | Source: Credit Suisse 2013. |
| After-tax Cost of Equity (6) | 13.6% | (6) = (1) + (2)*(5) |
| Debt Premium (7) | 3.3% | SolarCity term loan bears interest at an annual rate of LIBOR plus 3.25%. Source: SolarCity 2013 10k. |
| Pre-tax Cost of Debt (8) | 6.9% | = (1) + (7) |
| Nominal After-tax Weighted Cost of Capital (9) | 10.5% | = (1-(3))*(6)+(3)*(8)*(1-(4)) |
| Nominal Pre-tax Weighted Cost of Capital (10) | 10.5% | = (6) / (1-(4))*(1-(3))+(8)*(3) |

## Table A-5. Assumptions for "Commercial" Customer's Cost of Capital Using CAPM

| Variable | Value | Explanation |
|---|---|---|
| Risk-free Rate (1) | 3.6% | 20-Year Treasury Bill Constant Maturity. December 1, 2013. Source: FRED 2014. |
| Equity Beta (2) | 1.00 | Assumes an equity beta equal to the market as a whole. |
| Debt/Asset Ratio (3) | 50% | Source: Newell et al. 2012. |
| Tax Rate (4) | 40.2% | Assumes a federal corporate income tax rate of 35% and state corporate income tax rate of 8%. |
| Equity Risk Premium (5) | 5.3% | Source: Credit Suisse 2013. |
| After-tax Cost of Equity (6) | 8.9% | = (1) + (2)*(5) |
| Debt Premium (7) | 1.4% | = (8) - (1) |
| Pre-tax Cost of Debt (8) | 5.1% | Annual Moody's Seasoned Aaa Corporate Bond Yield, 2004- 2013. Source: FRED 2014. |
| Nominal After-tax Weighted Cost of Capital (9) | 6.0% | = (1-(3))*(6)+(3)*(8)*(1-(4)) |
| Nominal Pre-tax Weighted Cost of Capital (10) | 10.0% | = (6) / (1-(4))*(1-(3))+(8)*(3) |

## Table A-6. Assumptions for "Residential" Customer Cost of Capital

| Variable | Value | Explanation |
|---|---|---|
| Equity Percentage of Investment Portfolio (1) | 50% | Assumption |
| Debt Percentage of Investment Portfolio (2) | 50% | Assumption |
| Expected Equity Return (3) | 11.5% | Standard & Poors 500 IndexArithmetic Annual Average Return on Investment, 1928-2013. Source: FRED, 2014. |
| Expected Debt Return (4) | 5.9% | Moody's Seasoned Aaa Corporate Bond Arithmetic Annual Average Yield, 1928-2013. Source: FRED, 2014. |
| Capital Gains Tax (5) | 28.7% | U.S. average combined state & federal top marginal tax rate on capital gains. Source: Pomerleau & Borean 2014. |
| Nominal Pre-tax Weighted Cost of Capital (6) | 8.7% | (6)=(1)*(3)+(2)*(4) |

| Variable | Value | Explanation |
|---|---|---|
| Nominal After-tax Weighted Cost of Capital (7) | 6.2% | (7)=(6)*(1-(5)) |

# REFERENCES


Admirals Bank Home Improvement Lending. (2014). "Title I Secured & Unsecured Rate Matrix." Effective Date January 1, 2014.

Arizona Goes Solar. (2014). Information collected from the utilities Arizona Public Service and Salt River Project. Accessed April 7, 2014: http://arizonagoessolar.org/UtilityIncentives/ArizonaPublicService.aspx; http://arizonagoessolar.org/UtilityIncentives/SaltRiverProject.aspx.

Bankrate. (2014). "Checking and Savings." Accessed September 15, 2014: http://www.bankrate.com/checking.aspx.

Barbose, G. (2013). "State RPS Policies and Solar Energy Impacts, Experiences, Challenges, and Lessons Learned." Lawrence Berkeley National Laboratory. SEIA Webinar, November 21, 2013. http://emp.lbl.gov/sites/all/files/seia-webinar-nov-2013.pdf

Barbose, G.; Darghouth, N.; Wiser, R. (2011). *Tracking the Sun IV: An Historical Summary of the Installed Cost of Photovoltaics in the United States from 1998 to 2010*. Berkeley, CA: Lawrence Berkeley National Laboratory. Accessed September 2014: http://eetd.lbl.gov/ea/emp/reports/lbnl-5047e.pdf

Bolinger, M. (2014). *An Analysis of the Costs, Benefits, and Implications of Different Approaches to Capturing the Value of Renewable Energy Tax Incentives*. Berkeley, CA: Lawrence Berkeley National Laboratory. Accessed September 2014: http://emp.lbl.gov/sites/all/files/lbnl-6610e.pdf

California Solar Initiative (CSI). (2014). "Current CSI Data." Accessed April 4, 2014: www.californiasolarstatistics.ca.gov/current data files/.

Congress of the United States Congressional Budget Office (CBO). (2013). *The Distribution of Household Income and Federal Taxes 2010.*

Credit Suisse Research Institute. (2013). "Credit Suisse Global Investment Returns Yearbook 2013."

Desmarais, L. (2013). *The Impact of Photovoltaic Systems on Market Value and Marketability: A Case Study of 30 Single-Family Homes in the North and Northwest Denver Metro Area*. Denver, CO: Colorado Energy Office. Accessed September 3, 2014: http://www.colorado.gov/cs/Satellite?blobcol=urldata&blobheadername1=Content-Disposition&blobheader

name2=Content-Type&blobheadervalue1=inline;+filename%3D%22PV_ Case+Studies.pdf%22&blobheadervalue2= application/pdf&blobkey=id &blobtable=MungoBlobs&blobwhere=1251900073057&ssbinary=true.

FDIC (2012). *FDIC Community Banking Study*. Accessed January 6, 2014: https://www.fdic.gov/regulations/resources/cbi/study.html.

Federal Housing Finance Agency (FHFA). (July 6, 2010). *FHFA Statement on Certain Energy Retrofit Loan Programs.*

Federal Reserve Economic Data (FRED). (2014). Economic Research Division. Federal Reserve Bank of St. Louis. http://research. stlouisfed.org/fred2

Feldman D.; Barbose, G.; Margolis, R.; Darghouth, N.; James, T.; Weaver, S.; Goodrich, A.; Wiser. R. (2014a). *Photovoltaic System Pricing Trends: Historical, Recent, and Near-Term Projections. 2014 Edition*. NREL/PR-6A20-62558. Golden, CO: National Renewable Energy Laboratory.

Feldman D.; Friedman, B.; Margolis, R. (2013b). *Financing, Overhead, and Profit: An In-Depth Discussion of Costs Associated with Third-Party Financing of Residential and Commercial Photovoltaic Systems*. NREL/TP-6A20-60401. Golden, CO: National Renewable Energy Laboratory.

Gup, B. (2011). *Banking and Financial Institutions: A Guide for Directors, Investors, and Borrowers*. Hoboken, NJ: Wiley & Sons Ltd.

Heffernan, S. (2005). *Modern Banking*. West Sussex, England: Wiley & Sons Ltd.

Hoen, B.; Klise, G.; Graff-Zivin, J.; Thayer, M.; Seel, J.; Wiser, R. (2013). *Exploring California PV Home Premiums*. Berkeley, CA: Lawrence Berkeley National Laboratory. Accessed September 3, 2014: http://emp.lbl.gov/publications/exploring-california-pv-home-premiums.

Huynh, F. (February 3, 2014). "US Credit Quality Continues to Inch Forward." FICO Banking Analytics Blog. http://bankinganaly ticsblog. fico.com/2014/02/us-credit-quality-continues-toinch-forward. html.

Jickling, M.; Murphy, E. (2010). *Who Regulates Whom?An Overview of U.S. Financial Supervision*. The Congressional Research Service. Accessed March 18, 2014: http://fas.org/sgp/crs/misc/R40249.pdf.

Lee, B.; Daniels, T.; Boril, B. (2014). "Sell-off provides a better entry point – Remain Attractive on Solar." Goldman Sachs Equity Research.

Linder, S.; Lilinshtein, J.; Culver, N.; Di Capua, M.; Chase, J. (2014). *H1 2014 US PV Market Outlook*. Bloomberg New Energy Finance.

Litvak, N. (2014). *U.S. Residential Solar Financing: 2014-2018*. Boston, MA: GTM Research.

Lowder, T.; Mendelsohn, M. (2013). *The Potential of Securitization in Solar PV Finance*. NREL/TP-6A20-60230. Golden, CO: The National Renewable Energy Laboratory. Accessed July 31, 2014: http://www.nrel.gov/docs/fy14osti/60230.pdf

Martin, K. (February 2014). "Cost of Capital: 2014 Outlook." *Chadbourne & Parke Project Finance News: February 2014.*

Massachusetts Department of Energy Resources (DOER). (2014). "RPS Solar Carve-Out Qualified Units." Accessed March 26, 2014: http://www.mass.gov/eea/docs/doer/rps-aps/solarcarve-out-units.xlsx.

NCUA. (2014b). "Credit Union Online." Accessed March 21 2014: https://cuonline.ncua.gov/.

Newell, S.; Spees, K.; Pfeifenberger, J.; Mudge, R.; DeLucia, M.; Carlton, R. (2012). *ERCOT Investment Incentives and Resource Adequacy*. The Brattle Group.

Peterson, L. (2012). "Primary Tax Equity Finance Structures Common to the U.S. Domestic Solar Energy Industry: June 2012." Reznick Group.

Pomerleau, K.; Borean, R. (2014). "How High are Capital Gains Tax Rates in Your State?" Tax Foundation. http://taxfoundation.org/blog/how-high-are-capital-gains-tax-rates-your-state

San Diego Gas & Electric (SDG&E). (2014). "Total Residential Electric Rates." Accessed July 23, 2014: http://www.sdge.com/total-residential-electric-rates

Segal, L. M.; Sullivan, D.G. (1998). "Trends in Homeownership: Race, Demographics, and Income." *Economic Perspectives* Federal Reserve Bank of Chicago (22); pp. 53-72.

SEIA and GTM Research. (2014b). U.*S. Solar Market Insight Report: Q2 2014*. Washington, D.C.: Solar Energy Industry Association.

Shah, V.; Booream-Phelps, J.; Min, S. (2013). "Distributed Generation to Herald New US Growth Era." Deutsche Bank Market Research.

Solar Energy Industry Association (SEIA) and GTM Research. (2014a). *U.S. Solar Market Insight Report: 2013 Year-In-Review*. Washington, D.C.: Solar Energy Industry Association.

Standard & Poor's Ratings Services (S&P). (2013). "SolarCity LMC Series I LLC (Series 2013- 1)."

The American Banker's Association (ABA). (2014). *The Business of Banking: What Every Policy Maker Needs to Know*. Accessed July 25, 2014: http://www.aba.com/Tools/Economic/Documents/Businessofbanking.pdf.

The Fed. (2014b). *Financial Accounts of the United States*. Federal Reserve Statistical Release. Accessed June 10, 2014: http://www.federal reserve.gov/releases/z1/current/accessible/f6.htm.

The National Credit Union Administration (NCUA). (2014a). "Credit Union and Bank Rates". Accessed March 21, 2014: http://www.ncua.gov/Data Apps/Pages/CUBNKMain.aspx.

The United States Federal Reserve System (The Fed). (2014a). *Holding Companies with Assets Greater Than $10 Billion*. Accessed February 18, 2014: https://www.ffiec.gov/nicpubweb/nicweb/top50form.aspx.

U.S Department of Energy (DOE). (2012). *SunShot Vision Study*. DOE/GO-102012-3037. Washington, DC: U.S. Department of Energy. Accessed April 2014: http://www1.eere.energy.gov/solar/pdfs/47927.pdf

Wade, W. (2014). "Rooftop Solar Leases Scaring Buyers When Homeowners Sell." *Bloomberg*. Accessed June 25, 2014: http://www.bloomberg.com/news/2014-06-23/rooftop-solar-leasesscaring-buyers-when-homeowners-sell.html.

Wesoff, E. (2014). *SolarCity Quietly Moving to Loans and Customer-Owned Solar*. Greentech Media. Accessed August 1, 2014: http://www.greentech media.com/articles/read/SolarCityQuietly-Moving-to-Loans-and-Customer-Owned-Solar.

## End Notes

[1] Some loan products, such as the New York State Energy Research and Development Authority's (NYSERDA) on-bill program, can be designed to finance both solar installations and energy efficiency upgrades.

[2] Loan programs launched during this period include those offered by: Admirals Bank; Digital Credit Union (in partnership with SunPower and more recently Sungage); Dividend Solar; Lightstream; Mosaic; OneRoof Energy; SolarCity; Sungage Financial; WJ Bradley (in partnership with SunEdison); and others.

[3] Third-party ownership has been hampered in some states or municipalities due to various regulations, including: only allowing regulated utilities to sell electricity (e.g., FL, NC, KY, OK); only offering incentives to the utility customer or homeowner (e.g., NY); or potentially charging property taxes on third-party installed systems (e.g., AZ). However, many of these obstacles have been overcome by offering a slightly different product (leasing to a customer, rather than a PPA) or a change in law. Additionally, most of the states with favorable solar laws also have laws favorable to third-party system ownership.

[4] There are other reasons a company or business may choose to use TPO to finance a solar system, which is discussed in more detail in Section 5. Additionally, because the 25D personal tax credit expires in 2017, while the section 48 credit merely drops to 10%, certain

projects may find it more favorable to finance solar through TPO with a 10% tax credit than own it themselves without any tax credit.

[5] As of October 2014.

[6] There are currently several popular structures to financing a tax-equity investment in a solar asset; these include a sale-leaseback, a partnership flip, and an inverted lease. The choice between these structures is in large part affected by the preferences of the tax-equity provider and the needs and ability of the developer or sponsor. The sale-leaseback structure was chosen for this analysis because of its prevalence in the marketplace; however, a different financing structure would not affect the general outcome of this analysis assuming the same financial hurdle rates apply.

[7] In capital leases, the lessee effectively owns the asset under tax and accounting rules.

[8] SDG&E currently has a four-tiered rate structure, with electricity rates increasing as more energy is consumed. Comparison retail rates were based on the average of SDG&E tier 2 and tier 3 residential rates for both summer and winter, as solar systems are typically built to displace electricity charged at higher tiers (SDG&E 2014). However, individual customer electricity consumption and solar production varies.

[9] There may be additional non-economic reasons for an individual choosing a higher priced product; while some government supported loan programs offer lower interest rates, their success has been hampered by the additional paperwork and requirements.

[10] System owners also potentially have an easier time selling their home versus one with a PPA or lease, which would require the buyer to agree to assume the contract and the third-party provider to agree to the new customer, which can be at risk if the customer does not have a minimum credit score. However, a PPA customer could always buy-out the contract, which is very similar to the loan prepayment a system owner would face in the event of a home sale. Additionally, as of October, 2013, SolarCity had successfully reassigned approximately 2% of their contracts due to the normal sale of a house (S&P, 2013).

[11] As a point of reference, most module warranties have 25-year terms.

[12] Any unused portion of the personal tax credit can be carried forward to the following taxable year until 2016 (it is unclear whether the tax credit can be carried forward after 2016), while the business tax credit can be carried forward 20 years.

[13] It should be noted that, like credit-score, there is a positive correlation between household income and the percent likelihood of homeownership. Therefore, while the lower quintiles may not pay enough in federal taxes to use a 30% ITC in year one, they are also less likely to own their own home (Segal & Sullivan 1998).

[14] The additional perceived risk in a loan could translate into a higher discount rate by the customer than with a TPO system (i.e., the customer will want a higher return for owning rather than leasing the system). A separate analysis was performed to look at the impact the change in customer's discount rate had on the model. While there was an increase in LCOE, particularly for loans with shorter terms (due to the increased cash outlay by the customer of principal and interest payments), the higher discount rates did not change the analysis significantly. This is due to the fact that the loan is the primary source of financing host-owned projects in this analysis and therefore the yield of the loan has a much larger impact on the cost of energy than the cost of capital for the customer.

[15] There is also a risk of a change in rate structures, such as net metering. However this is mostly a risk that the holder of a PPA and a system owner share.

# INDEX

## D



## E



## F

## G



## H



## I



## J



## L

## S



## T